THE HOME DAIRY

Milk, Butter, Yogurt, and Simple Cheese From a Family Cow or Goat

Hudson Briggs

The Homestead Starter Series — Book 22

Home Dairy, Raw Milk, and Animal Safety Disclaimer

This book provides general educational information about household dairy animal care, milk handling, butter, yogurt, and simple cheese. It is not medical, veterinary, legal, food-safety, commercial dairy, or professional advice.

Raw milk can contain dangerous pathogens, including Listeria, Salmonella, E. coli, Campylobacter, and Brucella. The CDC and FDA advise against consuming raw milk. Pasteurize milk for pregnant people, children under five, adults over sixty-five, immunocompromised people, guests, and anyone whose risk you do not know. Chill milk below 40 degrees F quickly, discard questionable milk or cheese, follow antibiotic withdrawal periods exactly, work with a veterinarian, and verify all local and state laws before selling or sharing dairy products.

Introduction: A Gallon a Day, Every Day

The first thing to understand about a home dairy is that it does not pause.

A vegetable garden can be neglected for a week in July and forgive you. Chickens can be left with a big feeder and a waterer for a long weekend. Bees do not care whether you are home. A milking animal has to be milked twice a day, roughly twelve hours apart, every day, for about ten months of the year - in the rain, on Christmas morning, with a fever, the day after a funeral.

That is the entire deal, and it is why most people who try a home dairy quit and why the ones who stay love it as much as they do.

What you get for it

Milk you actually control. You know what the animal ate, what she was treated with, how clean the pail was, and how fast it got cold. That is not a small thing, and it is genuinely unavailable at any price in a store.

Cream. Real cream, thick enough to stand a spoon in, from a Jersey cow. This is what makes people fall in love with a family cow.

Butter, yogurt, and cheese made from milk that was in the animal an hour ago. Fresh chevre made the same day is a different food from what you buy.

A closed loop. Whey goes to the pigs and chickens. Manure goes to the garden. The garden feeds you. Book 17 in this series covers what that manure does to your soil.

Skill. Milking, handling, and cheesemaking are old competencies, and having them changes how you think about food.

What it costs

Time, primarily. Milking, cleaning up, and animal chores run roughly ninety minutes to two hours a day. Doubled if you keep more than one or two animals in milk.

Money, more than you expect and less than you fear. A dairy goat runs two to six hundred dollars, a good family cow twelve hundred to thirty-five hundred. Feed, hay, minerals, hoof trimming, breeding, and veterinary care add a steady monthly cost. Equipment for milking and cheesemaking is a few hundred dollars up front.

Freedom. This is the one people underestimate. You cannot leave. Not for a weekend, not overnight, unless you have someone trained and willing to milk. Solving this problem - before you buy an animal - is the most important preparation in this book.

Will it save money? Almost certainly not, if you count your time at anything. Do this because you want the milk and the life around it, not because the arithmetic works.

What this book covers

Choosing between a cow and a goat and choosing a good individual animal. Housing, fencing, and feed. The lactation cycle and what it means for your year. Milking technique, hygiene, and udder health. Handling milk so it tastes good and keeps. The raw milk question, treated honestly. And then the kitchen: cream, butter, yogurt, kefir, soft cheeses, mozzarella, and a first aged cheese.

It ends with the legal side, breeding and drying off, and a first-year plan.

What it does not cover

Advanced cheesemaking. Chapter 19 gets you through a first aged cheese. Beyond that you want a dedicated cheesemaking book, and Chapter 19 names the good ones.

General livestock husbandry beyond what a dairy animal specifically needs. Book 7 in this series covers small livestock more broadly.

Meat processing. Book 11.

Commercial dairying. Entirely different regulations, scale, and economics.

One decision to make now

Before you read another chapter, answer this: who milks when you cannot?

An illness, a family emergency, a wedding out of state. You need at least one other person who can walk into that barn and do the job correctly. Ideally you train them before you need them, by having them milk alongside you a dozen times.

The homesteaders who last are the ones who solved this in advance. The ones who sell the cow in year two are usually the ones who did not.

Sort that out, and the rest of this is a pleasure.

PART ONE - BEFORE YOU COMMIT

Chapter 1: The Honest Commitment

Everything in this chapter is the part people skip.

The daily schedule

Twice a day, twelve hours apart. Six and six, five and five, seven and seven. Consistency matters - an animal's udder fills on a schedule and she becomes uncomfortable and less productive when you are erratic.

Per session, realistically:

The daily schedule

Task	Time
Setup, catch animal, prep	5-10 min
Udder prep and strip check	3-5 min
Milking - goat	10-15 min
Milking - cow	15-30 min
Strain and chill	5 min
Clean equipment	10-15 min
Total per session	35-60 min

Plus daily animal chores - feed, water, hay, muck out, check on everyone: another 20 to 40 minutes.

A realistic daily total: 90 minutes to 2 hours.

Once-a-day milking

A legitimate option that a lot of homesteaders use, and it deserves more attention than it usually gets.

How it works: milk once every 24 hours instead of twice every 12.

What it costs: roughly 20 to 40 percent less milk. The udder signals back that less is needed and production adjusts.

What it buys: half the labor, and a schedule you can actually live with.

How to transition: gradually, over a week or two, so the udder is not suddenly overfull. Watch closely for mastitis during the change.

For most families, once-a-day is the right answer unless you specifically need the volume. One goat milked once a day still gives you two to three quarts, which is plenty for a household that is not making cheese in bulk.

Calf and kid sharing

The other flexibility strategy, and an old one.

How it works: leave the calf or kid on the mother. Separate them overnight - twelve hours or so - then milk in the morning and let them back together for the day.

What it buys:

- You get one milking a day instead of two
- The offspring does the rest, which means no bottle feeding
- If you need to leave for a day, you simply do not separate them and the calf handles it. This is the single best solution to the vacation problem
- The udder gets emptied thoroughly, which is good for udder health

What it costs:

- Less milk for you
- You have to manage a separation pen and a gate
- The offspring gets bigger and hungrier over time, and eventually takes most of it

- Some animals will not let down for you if the calf is nearby

Worth serious consideration for a first-time dairy keeper. It reduces both the labor and the pressure.

The vacation problem

Solve this before you buy an animal.

Options:

A trained backup. A neighbor, family member, or fellow homesteader who can milk. Train them by having them do it alongside you repeatedly, until they are competent and the animal is comfortable with them. Then have them do a real session while you are gone for a day, before you need them for a week.

Calf sharing, as above. The most reliable answer.

Dry period timing. An animal is dry - not milking - for about two months a year. Schedule your travel then. This takes planning a year out and it works.

Boarding. Rare, and possible in areas with an active homesteading community.

"We just don't go anywhere." Some people genuinely choose this. Know if you are one of them.

Weather and conditions

You will milk in a January ice storm. You will milk when it is ninety-five degrees and the flies are bad. You will milk in the dark, at both ends of the day, for months.

A covered, lit, comfortable milking area is not a luxury. It is the difference between a chore you do reliably and one you start resenting. Chapter 4.

The animal is a living thing

She will get sick. You will call a vet, sometimes at inconvenient hours, and it will cost money.

She will get bred, and give birth. Usually uneventfully. Sometimes not, and a difficult birth at two in the morning is part of what you have signed up for.

She will produce offspring every year. You need a plan for what happens to them - kept, sold, or raised for meat. A dairy goat producing twins annually becomes a real management question by year three.

She will eventually stop being productive. Then you decide what to do with an animal you have milked twice a day for eight years. People find this harder than they expect.

Who this suits

It suits you if: you like routine, you are home most days, you enjoy animals, you want the milk badly enough to reorganize your life a little, and you have someone who can cover for you.

It does not suit you if: you travel often, your schedule is unpredictable, you want the products without the animal, or you are hoping to save money.

A middle path worth naming: buy raw milk from a local farm, if it is legal where you live, and make the butter, yogurt, and cheese in Part Five without owning an animal. You get most of the good part and none of the twice-a-day part. A lot of people should do this and never consider it.

Chapter 2: Cow or Goat?

The first real decision, and the one that shapes everything after it.

Milk volume

Milk volume

Animal	Daily production, in full lactation
Holstein cow	6-10 gallons
Jersey cow	3-6 gallons
Guernsey cow	3-5 gallons
Dexter cow	1.5-2.5 gallons
Saanen / Alpine goat	1-1.5 gallons
Nubian goat	0.75-1.25 gallons
Nigerian Dwarf goat	1-2 quarts

Start with your actual consumption. A family of four drinking a gallon a day, plus yogurt and some cheese, needs perhaps a gallon and a half daily.

A Jersey cow in full lactation will drown that family in milk. Four to six gallons a day is 30 to 45 gallons a week. You will be giving it away, feeding it to pigs, or making an enormous amount of cheese. This surprises people constantly.

Two good dairy goats produce roughly two gallons a day combined, which is much closer to what most households actually use.

Under-milked capacity is not free. You still feed the animal, still milk her twice a day, and still deal with the volume.

Butterfat, and what it means for cream

Butterfat, and what it means for cream

Animal	Butterfat
Jersey	4.5-5.5%
Guernsey	4.5-5%
Holstein	3.5%
Nubian goat	4.5-5%
Nigerian Dwarf goat	6-10%
Saanen / Alpine goat	3-3.5%

The critical difference is not the percentage. It is that cow milk separates and goat milk does not.

Cow milk: cream rises on its own. Set a jar in the refrigerator overnight and there is a distinct cream line you can ladle or siphon off. Butter, whipped cream, and heavy cream for cooking all follow directly.

Goat milk is naturally homogenized. The fat globules are smaller and stay suspended. Cream rises very slowly and incompletely, and what does rise is hard to skim cleanly. To get meaningful cream from goat milk you need a mechanical cream separator, which costs two to six hundred dollars.

If cream and butter are a priority, this points strongly toward a cow. If you mainly want drinking milk, yogurt, and cheese, goats are fine.

Cost and scale

Cost and scale

Method	Goat	Cow
Purchase price	$200-600 each	$1,200-3,500
Weight	100-200 lbs	800-1,400 lbs
Land needed	1/4-1/2 acre for a few	1-2+ acres
Feed cost per month	$25-50 each	$75-200
Handling difficulty	Easy - you can physically move one	You cannot move a cow that does not want to move
Fencing	Difficult - goats escape	Straightforward
Housing	Small shed	Real barn or shelter
Vet costs	Lower	Higher
Hoof trimming	Every 6-8 weeks, DIY	Less frequent, often professional

Goats are cheaper to buy, cheaper to keep, and easier to handle. A person of ordinary strength can physically manage a goat. Nobody manhandles a cow.

Cows produce more per unit of labor. One cow gives more milk than four goats, for one milking session instead of four.

The things nobody mentions

Goats must be kept in groups. A goat alone is a miserable, noisy, escape-prone animal. Minimum two, always. That doubles your purchase, your feed, and your milking labor from the start.

Cows also prefer company but tolerate solitude better, especially with other livestock around.

Goats escape. This is not a joke and it is the single most common complaint from new goat keepers. They climb, they push, they test every inch of fence, and they will find the weak point. Fencing a goat properly costs more than fencing a cow. Book 23 in this series covers fencing.

Goat milk flavor. Well-handled goat milk from a healthy doe tastes clean, slightly sweet, and mild - most people cannot distinguish it from cow milk in a blind test. The "goaty" taste comes from a buck kept near the does, poor sanitation, slow chilling, or an animal with health problems. It is a management failure, not an inherent property. That said, some people do detect a difference and dislike it. Taste the milk from the actual animal before you buy.

Cow milk fat globules are larger, and some people who struggle to digest cow milk do better with goat. This is individual, not universal.

Which to choose

Choose goats if: you have limited land, a limited budget, you want a manageable volume, you cannot handle a large animal, you want easier handling, or you want to start smaller and learn.

Choose a cow if: you want cream and butter, you want the most milk per hour of labor, you have the land and a barn, you can handle a large animal, and you want one animal instead of a herd.

Choose a Dexter or a mini-Jersey if you want cow milk at a manageable scale. A Dexter gives one and a half to two and a half gallons a day at half the size of a full Jersey. For many homesteads this is the honest sweet spot.

Most first-time home dairies should start with two dairy goats. Lower cost, lower risk, easier to handle, and if it turns out you hate milking twice a day you are out six hundred dollars instead of three thousand.

Sheep, briefly

Dairy sheep exist and produce excellent milk - very high in fat and solids, and the basis of some of the world's best cheeses. Their lactation is short, volume is low, and good dairy stock is hard to find in North America. Not a beginner path.

Chapter 3: Choosing Your Animal

The individual matters more than the breed.

Buy in milk, if you can

"In milk" means she is currently lactating and you can see exactly what she produces.

The alternative is buying a bred heifer or doeling and waiting - sometimes a year or more - to find out what you have. That is a gamble, and it is a poor one for a first animal.

Buying in milk lets you:

- Watch her being milked
- Milk her yourself, before you buy
- Taste the milk
- Measure actual production
- See her temperament in the situation that matters
- Check her udder and teats

Do all of those things. A seller who will not let you milk the animal is telling you something.

What to look for

Temperament, first. This matters more than production, more than pedigree, more than anything. You will spend an hour a day with this animal, in close physical contact, for years.

- Does she stand quietly on the stand?
- Does she kick, stamp, or swing at you?

- Is she easy to catch and lead?
- Does she come to people or avoid them?

A high-producing animal with a bad attitude is worse than a moderate producer who stands still. Walk away from kickers. There are plenty of gentle animals.

Udder and teats.

- Well attached, held up close to the body. A pendulous udder that swings low is prone to injury and mastitis, and it gets worse with each lactation
- Even quarters or halves - no one section noticeably larger or smaller, which can indicate past mastitis damage
- Soft and pliable after milking, not hard or lumpy
- Teats sized for your hands. This is genuinely important and rarely mentioned. Very small teats are hard to hand milk. Very large ones are awkward. Try before you buy
- Teats pointing down, not out to the sides
- No scarring, warts, or lesions

Body condition. Not thin, not obese. Bright eyes, clean coat, alert posture. Good feet with no limping.

Age. A second- or third-lactation animal is often the sweet spot - she has proven she can produce and settle, and she has years left. A first-freshener is an unknown. An older animal may be a bargain and may be near the end.

Health records. Vaccination history, testing, treatments, breeding dates, production records if kept.

Testing

Ask what the animal and the herd have been tested for. Requirements and norms vary by region - talk to your veterinarian about what matters where you are.

For cattle, brucellosis and tuberculosis are the historically serious ones, and Johne's disease is worth asking about.

For goats, the standard three are CAE (caprine arthritis encephalitis), CL (caseous lymphadenitis), and Johne's. All three are chronic, incurable, and can be brought onto your placc on a single animal.

Buy from a tested herd if you can. If the seller has never tested and cannot tell you, that is a meaningful risk - particularly for CL, which spreads readily and never leaves a property.

Look for lumps at the lymph node sites - under the jaw, in front of the shoulder, at the flank - which can indicate CL.

Where to buy

Directly from a small dairy or homestead. The best source. You see the conditions, the parents, and the management. The seller usually cares where the animal goes and will answer questions for years.

Breed associations and registries list breeders and often have classifieds.

Local homesteading and 4-H networks.

Livestock auctions: avoid. Animals at auction are frequently there because someone had a problem with them. No history, no testing, no recourse, and disease exposure from every other animal in the barn. It is not where a first dairy animal should come from.

Online classifieds: possible, with in-person inspection. Never buy sight unseen.

Questions to ask

- How much does she give per day, at what stage of lactation?
- How many times has she freshened?
- When was she last bred, and when is she due?
- Has she ever had mastitis? Which quarter?
- What has she been tested for, and when?
- What is she being fed?
- Is she used to a milking stand? A machine or hand milking?
- Has she ever had trouble giving birth?
- Why are you selling her?

That last question matters. Legitimate answers exist - downsizing, moving, too much milk, keeping her daughter instead. "She's a great animal" with no reason is worth probing.

Registered or not

Registration papers matter if you plan to sell breeding stock or offspring at a premium. Registered animals cost more and sell for more.

They do not matter if you want milk for your family. An unregistered grade doe from good stock will fill your pail exactly as well.

What matters more than papers: the animal's own udder, temperament, and production, and the same qualities in her dam.

How many

Goats: two minimum. Not negotiable - goats are herd animals and a single goat is genuinely unhappy.

A useful arrangement: two does, bred to freshen a few months apart, so one is in milk while the other is dry. That gives you year-round milk without ever having two animals in full production at once.

Cows: one is workable, especially with other livestock for company. Two if you want continuity across dry periods.

Start with the minimum. You can always add. Selling animals you regret buying is harder.

Bringing her home

Quarantine new animals for two to four weeks, separate from any existing stock, before introducing them. Watch for illness, and get any testing done during that window.

Expect a production drop. A move is stressful, and milk yield commonly falls for a week or two. It comes back.

Keep her feed the same for the first week or two, then transition gradually to what you will feed. Sudden feed changes cause digestive upset, which in a ruminant is a serious matter.

Expect her to be uncooperative at first. New place, new person, new routine. Be calm, be consistent, be patient. Most animals settle within two weeks.

Chapter 4: Housing, Fencing, and Feed

The infrastructure. Get it built before the animal arrives.

Shelter

What an animal genuinely needs: protection from wind, rain, and snow; shade in summer; a dry place to lie down; and ventilation.

What she does not need: heat, insulation, or a barn worth more than she is. Healthy dairy animals tolerate cold well. Damp and drafty is what harms them, not cold.

Goats: a three-sided shed facing away from prevailing wind is adequate in most climates. Fifteen to twenty square feet per goat inside. Goats hate rain intensely and will crowd into shelter at the first drop.

Cows: a run-in shed or a barn stall. Forty to sixty square feet per animal under cover, more if they are housed rather than on pasture.

Bedding: straw, wood shavings, or a deep litter system. Keep it dry. Wet bedding is where mastitis-causing bacteria live, and a dirty udder is the direct route from the bedding into the pail.

Ventilation without drafts. Openings high, solid walls low. Ammonia smell means inadequate ventilation and it damages lungs.

The milking area

This deserves more thought than most people give it, because you will use it seven hundred times a year.

It should be:

- Covered and dry
- Lit. You will milk in the dark for months. A good light is transformative
- Clean, and cleanable. A hard floor you can sweep or hose. Not deep bedding
- Separate from the sleeping and manure area if possible
- At a comfortable working height - see the stand below
- Set up with everything within reach: pail, wash supplies, teat dip, a place to set the strainer
- Fly controlled in summer. Flies make an animal fidget, and a fidgeting animal in a milking stand is unpleasant and sometimes dangerous

A milking stand for goats is close to essential - a raised platform with a head stanchion and a grain bucket. It puts the udder at a workable height, holds her still, and the grain gives her a reason to want to be there. Plans are widely available and it is a straightforward weekend build. Book 6 in this series covers projects of this kind.

For cows, a stanchion or a simple headgate with a feed trough, plus a stool. Some people milk a well-trained cow standing loose in a stall; a headgate is safer and more reliable.

Fencing

For cows: relatively simple. Woven wire, high-tensile electric, or board fencing. Four strands of properly grounded electric will hold most cattle once trained to it.

For goats: a genuinely difficult problem.

Goats climb, push, jump, squeeze, and lean on fences until they fail. They will find the one weak spot in a quarter mile of fence and they will find it the first day.

What works:

- Woven wire field fence, at least four feet, with the smaller square sizes. Watch for horned goats putting their heads through and getting stuck
- Electric netting, portable and effective, and it must stay hot. A goat that learns the fence is off will never respect it again
- High-tensile electric, five or more strands, with the lowest close to the ground
- Cattle panels - extremely effective, expensive over long runs, excellent for small pens

What does not work: barbed wire, which goats ignore and injure themselves on. Anything with a gap a head fits through. Anything they can climb, which includes stacked pallets, feeders near the fence, and anything else you leave next to it.

Test it by walking the line and pushing on it. Where you can move it, they will move it.

Book 23 in this series covers fencing and gates in detail.

Feed

Ruminants live on forage. Grain is a supplement, not the foundation.

Forage - the base of the diet:

Pasture, when available. The cheapest and best feed there is.

Hay, the rest of the year. Quality matters enormously:

- Grass hay - timothy, orchard grass, brome. The staple
- Legume hay - alfalfa, clover. Higher protein and calcium, well suited to a milking animal, and it costs more

- Good hay is green, leafy, sweet-smelling, and free of dust and mold. Brown, stemmy, dusty, or musty hay is poor feed and moldy hay is genuinely dangerous

How much:

- Goat: 3 to 5 pounds of hay per day, or free-choice
- Cow: 25 to 40 pounds of hay per day, or free-choice

Free-choice hay is the simplest approach - keep the feeder full and let them eat what they need. Use a feeder that keeps hay off the ground, because forage contaminated with manure is both wasted and a parasite route.

Grain - the supplement:

Milking takes energy, and a high producer often cannot eat enough forage to cover it.

- Goat: roughly 1 pound of grain per 3 pounds of milk produced, commonly 1 to 3 pounds a day
- Cow: commonly 4 to 12 pounds a day depending on production

Feed grain on the milking stand. It gives the animal a reason to cooperate and it makes the whole routine easier.

Do not overfeed grain. Excess grain in a ruminant causes acidosis, bloat, and in bad cases death. Increase gradually, never suddenly.

Minerals - the part people neglect:

Free-choice loose minerals formulated for the species. Not a block - a block is for cattle tongues and goats cannot get enough off one.

Goat minerals must contain copper. Sheep minerals are formulated without it, because copper is toxic to sheep. Never feed sheep minerals to goats - copper deficiency causes a rough coat, a faded coat color, fish-tail hair loss, poor growth, and reproductive failure.

Selenium is deficient in soils across large regions of North America. Ask your extension office or vet whether yours is a deficient area and supplement accordingly.

Calcium and phosphorus balance matters, particularly for males, where an imbalance causes urinary calculi.

Baking soda free choice helps buffer rumen pH.

Water:

Clean, unlimited, year round. A milking animal drinks a great deal - a lactating goat may drink two to four gallons a day, a cow twenty to thirty or more.

Water intake drives milk production directly. An animal short on water gives less milk, immediately.

Heated waterers or tank heaters in winter. Breaking ice twice a day gets old in January and animals drink less when the water is near freezing.

Startup infrastructure budget

Startup infrastructure budget

Item	Cost
Shelter, if building simple	$200-1,500
Fencing	$300-2,000+
Milking stand (goat) or stanchion	$50-200
Hay feeder	$50-200
Water tank and heater	$75-200
Mineral feeder	$20
Milking equipment (Ch. 8)	$150-300

Have it all built and installed before the animal arrives. Improvising a fence with a loose goat in the yard is a bad first day.

PART TWO - KEEPING A MILKING ANIMAL

Chapter 5: The Lactation Cycle

Milk is a byproduct of birth. Understanding that cycle is understanding your year.

The basic fact

An animal produces milk because she had a baby. No pregnancy, no milk. To keep milk coming, you breed her roughly once a year, forever.

This is the part of home dairying that people gloss over and then run into. Milk means annual breeding, annual birth, and annual offspring you must do something with.

The cycle

Freshening. Giving birth and beginning lactation. A cow calves; a goat kids.

Colostrum, days 1 to 3. The first milk - thick, yellow, and packed with antibodies. This belongs to the newborn. A calf or kid that does not get adequate colostrum in the first hours has a poor chance. Do not take it for the house. Excess can be frozen for future newborns.

Peak production, weeks 4 to 8. Output climbs steeply after freshening and peaks somewhere in the second month.

The plateau and decline. Production holds, then declines slowly across the rest of the lactation. By month eight or nine she may be giving half her peak.

Breeding back. A cow is typically bred 60 to 90 days after calving; a goat when her season comes around.

Drying off. Milking stops six to eight weeks before the next birth for a cow, about two months for a goat. She needs that time to rest the udder and put resources into the developing offspring.

And again.

Gestation and timing

Gestation and timing

Method	Gestation	Standard lactation	Dry period
Cow	~283 days (9 months)	~305 days	6-8 weeks
Goat	145-155 days (5 months)	~305 days	~2 months

The goat's short gestation is a real advantage for planning. Five months from breeding to milk, versus nine.

Planning your milk year

With one animal: you have milk for about ten months and none for two. Plan around it - freeze milk, make hard cheese to age through the dry period, or simply buy milk for two months.

With two animals staggered: breed them to freshen three to six months apart and you have milk continuously, and you rarely have two in full production at once. This is the arrangement most experienced homesteaders end up with, and it is worth planning from the beginning.

Choosing your freshening date lets you decide when the work falls:

- Spring freshening matches peak production to peak pasture, which is cheapest. It also puts birthing in mud season
- Fall freshening gives you milk through winter and means birthing in cold, which requires more attention
- Avoid freshening in the worst weather of your year if you can. A newborn in an ice storm is a hard night

Seasonal breeders

Most goat breeds are seasonal breeders, cycling in fall as day length shortens, which means spring kids. Nigerian Dwarves and a few other breeds cycle year round, which gives you scheduling freedom.

Cattle cycle year round.

Heat cycles

Heat cycles

Method	Cycle length	Standing heat
Cow	~21 days	12-18 hours
Goat	18-21 days	12-36 hours

Signs of heat in a goat: tail flagging, vocalizing, restlessness, a swollen and reddened vulva, discharge, riding other does, and obsessive interest in a buck if one is anywhere nearby.

Signs in a cow: standing to be mounted - the definitive sign - restlessness, bellowing, clear stringy mucus, a swollen vulva, and a drop in milk production.

Missing a heat costs you 21 days. Watch closely, and write down every heat you observe. That record lets you predict the next one within a day or two.

Birth, briefly

Most births are uneventful and the animal handles it alone. Your job is mostly to be present and not to interfere.

Be ready with: clean towels, iodine for the navel, gloves, your veterinarian's number, and a plan for a hard night.

Normal presentation is front feet first with the nose resting on them. Anything else - a head with no feet, a tail first, one leg back - may need help, and help you are not trained to give means a phone call.

Call the vet if: hard straining for an hour with no progress, an obviously wrong presentation, the animal gives up and stops trying, or anything alarming. Have the number in your phone before you need it, and know whether your vet takes night calls, because many rural practices have shifted away from large animal work and finding out at 2 a.m. is too late.

After birth: make sure the newborn nurses within an hour or two. Dip the navel in iodine. Confirm the placenta passes - retained afterbirth beyond about 12 hours is a veterinary matter.

Milk fever - a calcium crash shortly after freshening, mostly in high-producing animals - presents as weakness, staggering, cold ears, and collapse. It is an emergency and it responds dramatically to prompt treatment. Know the signs.

What happens to the offspring

Plan this before breeding.

Females can be kept as replacements or sold. Good dairy stock sells readily.

Males are the hard part. Dairy bucklings and bull calves have little value as breeding stock and there are far more than anyone wants. Realistic options: raise for meat (Book 11), sell for meat or as pets, or give away. There are more male goats born every spring than there are homes.

Wethering - castrating a male - makes for a calmer animal that can be kept as a companion or raised for meat.

Be honest with yourself now. People who have not thought this through end up with a herd of aging wethers eating hay and doing nothing. It is one of the most common ways a small dairy operation stops making sense.

Chapter 6: Feeding for Milk

Milk is made from feed. What goes in determines what comes out, in volume and in flavor.

The ruminant

A cow or goat has four stomach compartments, and the large one - the rumen - is a fermentation vat full of microbes that break down cellulose the animal cannot digest itself.

Two consequences matter:

Forage is the foundation. The rumen microbes are adapted to fiber. An animal fed mostly grain has a disrupted rumen and gets sick.

Change feed slowly. The microbial population takes days to adapt. A sudden change - a new hay, a jump in grain, a burst of spring pasture - can cause bloat, acidosis, or diarrhea. Transition over seven to ten days, always.

Forage

Pasture is ideal when available: fresh, cheap, and it does good things for milk flavor and fat composition.

Hay is the year-round staple in most places.

Judging hay:

- Green, not brown or yellow
- Leafy, not all stem. The leaves carry the nutrition
- Sweet-smelling
- No dust, no mold, no mustiness. Moldy hay is dangerous. Do not feed it. Do not try to sort it
- Cut at the right stage - before heavy seed head development

Alfalfa or a grass-legume mix suits a milking animal well, providing the protein and calcium lactation demands.

Store hay dry and off the ground. Wet hay molds, and stacked wet hay can genuinely combust.

Grain

Grain supplies the energy that forage alone often cannot for a high producer.

Rules of thumb:

- Goat: about 1 pound of grain per 3 pounds of milk. Commonly 1 to 3 pounds a day
- Cow: commonly 4 to 12 pounds a day, scaled to production

A commercial dairy ration formulated for the species is the simple, reliable choice. It is balanced and it includes minerals.

Feed it on the stand. Grain is your leverage - she comes willingly, she stands still, and the routine works.

Overfeeding grain causes real problems: acidosis, bloat, laminitis, obesity, and reduced fertility. More is not better.

Increase gradually as production rises after freshening, and taper as production declines.

Minerals

Free-choice loose minerals, species-specific, available at all times.

For goats, the mineral must contain copper. Never feed sheep or all-stock minerals to goats - copper is toxic to sheep so those formulations omit it, and copper-deficient goats show a rough faded coat, fish-tail hair loss on the tail tip, anemia, poor growth, and infertility.

Selenium is deficient in soils across much of North America. Your extension office knows whether yours is one of those areas. Deficiency causes white muscle disease in newborns and reproductive failure in adults.

Calcium and phosphorus must stay in balance, roughly 2:1. Alfalfa is high in calcium; grain is high in phosphorus. An imbalance in males causes urinary stones, which are frequently fatal.

Baking soda free choice lets the animal self-regulate rumen acidity.

Kelp meal is a common homestead supplement providing trace minerals and iodine.

Water

The most underrated input. Milk is roughly 87 percent water.

- Lactating goat: 2 to 4 gallons a day
- Lactating cow: 20 to 30+ gallons a day

Clean, fresh, and unlimited. Scrub tanks regularly - animals drink less from fouled water and drinking less means giving less.

Warm it in winter. Consumption drops sharply with near-freezing water, and so does production.

Feed and milk flavor

This is where a lot of "bad milk" comes from, and it is almost always fixable.

Feeds that taint milk:

- Wild onion and wild garlic - the classic culprit. A pasture with wild garlic in spring produces milk nobody wants to drink

- Brassicas - turnips, kale, cabbage, rape - in quantity
- Some strong weeds - bitterweed, ragweed, chamomile
- Silage, if poorly fermented

Fix: keep the animal off the offending pasture, or move her off it several hours before milking. Flavor compounds pass through quickly and clear within a few hours.

Buck taint is not a feed problem and it is the most common cause of "goaty" milk. A buck in rut produces a powerful odor that permeates everything, including does housed near him, and it comes through in the milk.

Keep bucks well away from milking does - a separate pen at a distance, ideally out of the wind line, and certainly not in the milking area. Do not handle the buck and then milk without changing clothes and washing thoroughly.

Body condition

Score her regularly by running a hand over the ribs, spine, and hips.

Too thin: the udder is stealing from her body. She needs more feed, or she has parasites, or she is producing more than she can support. Very common in high-producing goats in early lactation.

Too fat: a fertility problem and a birthing problem. Fat animals have harder births and more metabolic disease.

Early lactation is when she loses condition - production peaks before intake catches up. Expect it, feed her well, and watch that she recovers by mid-lactation.

Feeding through the cycle

Early lactation, weeks 1-8. Highest demand. Best hay, grain increasing with production, free-choice minerals, unlimited water. Watch for condition loss.

Mid lactation. Steady. Adjust grain to actual production.

Late lactation. Production declining. Reduce grain. She should be regaining condition.

Dry period. Cut grain substantially or eliminate it. Good hay and minerals. Do not let her get fat, which causes birthing problems.

Last few weeks before freshening. Reintroduce a small amount of grain to prepare the rumen for the lactation ration. This is called steaming up, and it reduces the metabolic shock of freshening.

Chapter 7: Health, Mastitis, and the Vet

The medical side. Most of it is prevention.

Find a vet before you need one

Do this first. Large animal veterinarians are increasingly scarce in many regions, and discovering that nobody within sixty miles sees goats is a bad thing to learn during an emergency.

Ask:

- Do you see cattle? Goats? Both?
- Do you take after-hours emergency calls?
- Do you do farm visits, or must I haul the animal in?
- What do you charge for a farm call?

Establish the relationship with a routine visit before there is a crisis. Many vets will not take an emergency call from someone who is not an existing client.

Your local homesteading network knows who the good large animal vets are. Ask before you buy an animal.

Daily observation

The most valuable health tool you have is looking at the animal every day, which milking makes automatic.

What normal looks like: alert, eating, chewing cud, moving freely, bright eyes, clean nose, normal manure, in the group.

What to notice:

- Off feed, or not chewing cud
- Standing apart from the others

- Head down, ears drooping
- Limping
- Coughing, nasal discharge
- Scours (diarrhea) or unusual manure
- A sudden production drop
- Grinding teeth, which signals pain
- Straining
- Any swelling anywhere

A sudden drop in milk is often the first sign of illness, before anything else is visible. Track your daily yield and you will catch problems early.

Normal temperatures: roughly 101.5-103.5 degrees F for goats, 100.5-102.5 degrees F for cattle. Own a thermometer and know how to use it. A temperature is the single most useful piece of information you can give a vet on the phone.

Mastitis

The disease that matters most in a dairy animal. It is inflammation of the udder, usually bacterial, and it can destroy a quarter permanently.

Signs of clinical mastitis:

- Clots, flakes, or stringy material in the milk - usually the first sign, and why you strip into a cup every milking
- Watery, discolored, or bloody milk
- A hot, hard, swollen, or painful quarter or half
- The animal kicking during milking when she normally does not
- Off feed, fever, lethargy in severe cases

- A sharp drop in production

Subclinical mastitis has no visible signs and shows up as reduced production and elevated somatic cell count. A California Mastitis Test paddle - twenty dollars, takes a minute - detects it, and it is worth running monthly.

What to do:

1. Call your vet, especially for a first case or a severe one. Different bacteria require different treatment and guessing wastes time the udder does not have
2. Milk out that quarter frequently and completely - every few hours. Removing infected milk is the single most important physical treatment
3. Warm compresses and gentle massage before milking, to help let-down
4. Discard the milk from the affected quarter
5. Intramammary antibiotics, if your vet prescribes them
6. Observe the withdrawal period exactly. This is a legal requirement and a safety one. Do not drink it, do not make cheese with it, do not feed it to animals you will eat

Prevention, which is most of the game:

- Clean, dry bedding. Most mastitis bacteria come from the environment
- Clean, and critically, dry teats before milking. Water running down a wet teat carries bacteria straight to the opening
- Milk out completely every time. Residual milk is a growth medium
- Post-milking teat dip, every teat, every milking. The most effective single practice there is

- Keep her standing for 20 to 30 minutes after milking - the teat canal stays open that long, and an animal lying in bedding with open teats is how infection enters. Feeding hay after milking accomplishes this automatically
- Clean hands and clean equipment
- Treat at dry-off if she has a history - ask your vet about dry-period therapy
- Cull chronic cases. An animal with repeated mastitis is a source of infection for the others

Parasites

The leading health problem in goats, and a significant one in cattle.

Internal parasites - barber pole worm especially in goats - cause anemia, weight loss, bottle jaw (fluid swelling under the jaw), poor coat, and death.

FAMACHA scoring is a field method for goats and sheep: check the color of the inner lower eyelid against a printed chart. Pale means anemic means treat. Get trained - many extension offices run FAMACHA certification, and it lets you treat the animals that need it rather than the whole herd.

Fecal egg counts through your vet tell you what parasites you actually have and whether treatment worked.

Resistance is a serious and growing problem. Blanket deworming of every animal on a schedule is exactly how resistance develops. Treat individuals that need it, based on scoring or counts, not the whole herd on a calendar.

Pasture management matters more than drugs: rotate pasture, avoid overgrazing (parasite larvae concentrate in the bottom few inches of grass), keep feed and water off the ground, and rest paddocks.

External parasites - lice, mites, ticks, flies - cause itching, hair loss, and irritation. Treatable, and check withdrawal times for anything you apply to a milking animal.

Hooves

Trim goats every 6 to 8 weeks. Overgrown hooves cause lameness, deformity, and hoof rot. It is a straightforward DIY job with hoof shears; have someone show you once.

Cattle need less frequent attention, often professional.

Check feet at every milking. Lameness is one of the more common reasons an animal goes off feed and off production.

Vaccination

CD&T - clostridial enterotoxemia types C and D plus tetanus - is the standard core vaccine for goats and sheep. Annual boosters, and a pre-freshening dose so newborns get antibodies through colostrum.

Cattle vaccination programs vary widely by region and management.

Ask your vet what is appropriate where you are. Regional disease pressure varies a great deal.

Withdrawal times

Any medication given to a milking animal has a withdrawal period during which the milk must not be consumed.

Rules:

- Read the label. Follow it exactly

- Many products are not labeled for dairy animals at all, and using them requires veterinary direction and often extended withdrawal
- When in doubt, ask your vet and discard longer
- Mark the calendar the day you treat, and mark the day you can resume
- Do not feed withdrawal milk to animals you will eat

The medicine shelf

Reasonable to have on hand:

- Digital thermometer
- Iodine, for navels and general use
- Teat dip
- Syringes and needles, if you are comfortable giving injections
- Electrolytes
- Probiotic paste, for after any stress or antibiotic course
- Vet wrap and gauze
- Gloves, including shoulder-length for birthing
- Lubricant
- Hoof shears
- CMT paddle and solution
- FAMACHA card, if you have goats
- Your vet's number, written where anyone can find it

Know your limits. Giving a subcutaneous injection is a reasonable skill to learn. Diagnosing a sick ruminant from an internet forum is not. Ruminants hide illness well, and by the time a goat looks obviously sick she is often very sick. Call early rather than late.

PART THREE - MILKING

Chapter 8: Equipment and Sanitation

Milk quality is decided here, before a single drop is drawn.

The equipment

A stainless steel pail. Fifty to a hundred dollars. Seamless, food-grade stainless.

Why stainless: it does not react with milk, it does not scratch into bacterial harbors the way plastic does, and it sanitizes properly. Aluminum reacts with milk acids. Plastic scratches, and every scratch is a place bacteria live through washing. Galvanized metal is not food safe.

A seamless pail matters because seams trap milk residue that no amount of scrubbing reaches.

A hooded or half-covered pail - with a partial lid - keeps hair, dust, and the occasional swinging tail out. Worth the extra cost.

A strainer and filters. A stainless milk strainer with disposable filter discs, about thirty dollars plus filters. Milk filters are cheap and single use.

Storage jars. Half-gallon glass canning jars are ideal - nonreactive, sealable, and you can see the cream line. Wide mouth is easier to clean.

A thermometer. For pasteurizing and cheesemaking.

A strip cup. A small cup, ideally with a dark or screened surface, for the first squirts from each teat. Lets you see clots. A dark bowl works.

Udder wash. Warm water with a dairy-specific udder wash, or single-use udder wipes.

Paper towels. One per teat, or per quarter. Never a shared cloth, which moves bacteria from one teat to the next.

Teat dip. Iodine-based or chlorhexidine, in a dip cup. Ten to twenty dollars and it is the single most effective mastitis prevention available.

A hobble, optional, for an animal that kicks while you are training her.

Washing procedure

The order matters, and most people get it wrong.

1. Cold or lukewarm rinse first.

Immediately after milking. Never hot water first - heat cooks milk protein onto the surface and creates a film that is very hard to remove and that harbors bacteria.

2. Hot soapy wash.

Hot water with a dairy detergent or a good dish soap. Scrub every surface with a dedicated brush. Pay attention to seams, handles, the strainer, and anywhere milk pools.

3. Acid rinse, periodically.

A dilute acid rinse - vinegar in water, or a dairy acid cleaner - removes the mineral film ("milkstone") that builds up over time from hard water and milk minerals. Weekly is reasonable; more often with hard water.

4. Rinse clean.

5. Air dry, inverted, on a rack.

Do not towel dry. A towel puts back everything you just removed. Air drying on a rack is the standard, and it is also the easiest.

6. Sanitize before use, not after storage.

If you sanitize, do it right before milking rather than after washing. A sanitized surface that sits for twelve hours is not sanitized anymore.

Dedicated brushes. Keep milk equipment brushes separate from general dish brushes.

Chlorinated water

Chlorine in wash water is fine - it helps. But rinse thoroughly, because chlorine residue inhibits cheese cultures. If you make cheese, use non-chlorinated water for the final rinse and for cheesemaking itself.

Hygiene at milking

Wash your hands with soap before you start. Again if you touch anything else mid-session.

Wear clean clothes. Especially relevant if you have handled a buck, manure, or another animal.

Tie back long hair.

Clean the udder and flanks. Brush off loose dirt and hair before you start. A dusty animal drops dust into the pail.

Wash the teats, then dry them completely. Drying is not optional. Wet teats drip contaminated water straight into the pail, and every drop of that water was on the outside of the animal a moment ago.

Strip the first squirts into the strip cup, not the pail. This does two things: it lets you check for clots, and it clears the teat canal, which holds the highest bacterial count of any milk in the udder.

Keep the pail out from under until you are milking into it, and cover it or set it aside if you have to step away.

Watch for the tail. A tail through the pail undoes everything.

Flies

Flies make an animal fidget and they carry bacteria.

- Keep the milking area clean and manure removed
- Fly traps and sticky tape, away from the milking area itself
- Fans - flies avoid moving air, and a box fan in the milking area works remarkably well
- Fly predators, the parasitic wasps sold for manure management
- Animal-safe fly spray, checking any withdrawal statement

The single biggest quality factor

Speed of chilling. Chapter 11 covers it in full, and it belongs here too because it starts at the pail.

Milk leaves the animal at about 101 degrees F, which is an ideal bacterial incubation temperature. Every minute it sits warm, bacterial counts climb.

Get it strained and into a cold environment as fast as you physically can. Ideally within thirty minutes of the first squirt.

Milk that is chilled fast tastes clean and keeps a week or more. The same milk chilled slowly tastes off in three days. This is the difference people attribute to breed, feed, or luck, and it is almost always chilling.

Chapter 9: Hand Milking, Step by Step

The skill itself. It takes most people one to three weeks to become competent.

The mechanics

The thing everyone gets wrong at first: you do not pull.

Pulling down on a teat hurts the animal, damages the tissue, and does not get milk out. It is the reason a new milker gets kicked.

What actually happens:

1. Trap the milk in the teat by closing your thumb and index finger around the top, where the teat meets the udder. This seals the milk in the teat so it cannot flow back up

2. Squeeze the trapped milk downward and out by closing your remaining fingers in sequence - middle, then ring, then little finger - like a rolling wave

3. Release completely so the teat refills from the udder

4. Repeat

It is trap, squeeze, release. The hand stays in place. Nothing pulls.

Rhythm: alternate hands so one squeezes while the other releases. A steady two-handed rhythm is faster and easier on the animal than one hand at a time.

Before you start

1. Wash your hands

2. Get the animal on the stand or in the stanchion, with grain

3. Brush off loose dirt and hair

4. Wash the udder and teats with warm water and udder wash, or wipes

5. Dry completely. Paper towel, one per teat

6. Strip two or three squirts from each teat into the strip cup. Look at them. Clots, flakes, blood, or watery milk mean stop and see Chapter 7

Milking

7. Position the pail under the udder, out of the path of a foot

8. Start milking, two teats at a time. Cows have four quarters; work them in pairs

9. Find the rhythm. Slow and steady beats fast and rough. Speed comes with weeks, not with effort

10. Expect let-down. A minute or so in, the animal releases oxytocin and the udder firms noticeably as milk moves down. Flow increases sharply. This is why you should not rush the start - and why stress, pain, or a strange environment can prevent let-down entirely

11. Milk until the flow stops. The teats will feel soft and empty and the udder will be slack

12. Strip out the last of it. Massage the udder downward and milk again. The last milk is the highest in fat, and more importantly, milk left behind reduces future production and invites mastitis

After

13. Dip every teat in teat dip. Every teat, every milking

14. Keep her standing 20 to 30 minutes. The teat canal stays open that long. Give her hay after milking and she will do this herself

15. Strain and chill immediately. Chapter 11

16. Wash equipment. Chapter 8

Learning it

Your hands will hurt. Forearm cramps and hand fatigue are universal for the first week or two. It is grip endurance, and it builds. Squeezing a ball between sessions helps.

You will be slow. Thirty minutes for a goat that an experienced milker does in ten. Normal.

The animal knows you are new. She will fidget, step, and test you. Be calm and be consistent.

Get someone to show you once. Ten minutes of watching a competent milker teaches more than this chapter can. Ask the seller when you buy the animal - most will happily demonstrate.

Common problems

No milk coming out. You are pulling, or you are not sealing at the top. Check the thumb-and-index seal.

Milk squirting everywhere. Aim comes with practice. Everyone sprays their own leg for a while.

She kicks. Common causes: pain from a mastitis quarter or a chapped teat, cold hands, poor technique, a new milker, or flies. Investigate before assuming she is being difficult. Hobbles help while you train, and gentleness helps more.

She holds her milk up. No let-down. Causes: stress, pain, a strange environment, a buck nearby, or a calf she is saving it for. Be calm, feed grain, warm your hands, give it a minute. Persistent refusal to let down usually has a cause worth finding.

She lies down. Common in a young or untrained animal. Get her up, restart. A properly built stand makes this harder for her.

She puts a foot in the pail. Everyone loses milk this way. Position the pail off to one side and keep a hand ready to pull it clear.

Milk sprays back into your face. You are squeezing with a poor seal.

Training a first-freshener

An animal who has never been milked needs teaching.

- Start before she freshens. Put her on the stand daily with grain, and handle her udder. She learns the routine before it matters
- Be patient. Expect two to three weeks
- Use grain generously at first
- Hobbles, if she kicks, and only while you train
- Short sessions, ending on a calm note
- Never lose your temper. An animal that becomes afraid of the stand is a long project to undo

Chapter 10: Machines and Milking Problems

When hand milking stops being the right answer, and what to do when things go wrong.

Milking machines

When a machine makes sense:

- More than two or three animals in milk
- Hand or wrist problems - arthritis, carpal tunnel, injury
- Time constraints
- A large cow with high production

Types:

Bucket milker. A vacuum pump, a pulsator, a lidded bucket, and a claw with inflations. The standard small-scale unit. $600 to $2,000.

Portable single-unit milkers. Smaller and cheaper, from a few hundred dollars. Quality varies enormously - some are excellent and some are underpowered toys.

Pipeline systems. Commercial scale. Not relevant here.

What matters in a machine:

Pulsation. A machine must pulse - alternating vacuum and rest - to mimic the natural squeeze-and-release. Continuous vacuum damages teat tissue. Any machine without a working pulsator will injure your animal. This is the specification to check first.

Correct vacuum level for the species. Goats and cattle differ. Too high damages teats; too low milks incompletely.

Correct inflation size. The rubber liners must fit the teats. Wrong-sized inflations cause damage and slipping.

Cleanability. A machine is far harder to clean than a pail - hoses, a claw, inflations, and a pulsator all hold milk residue. This is where machine users create milk quality problems. Clean it fully, every time, on the same cold-rinse-then-hot-wash-then-acid schedule.

Replace inflations on schedule. Rubber degrades, cracks, and harbors bacteria. Follow the manufacturer's interval.

Downsides: cost, cleaning burden, noise, another thing to break, and it does not free you from being present.

A machine is not a beginner's first purchase. Learn to hand milk. You will need the skill anyway when the machine fails or the power goes out.

Production problems

Sudden drop in milk:

- Illness - check temperature first
- Mastitis
- Heat cycle, which commonly causes a temporary dip
- Stress: a move, a new animal, a dog, weather
- Feed change or a shortage of water
- Missed or delayed milking
- Pregnancy advancing

Gradual decline: normal across a lactation. Also check feed adequacy, parasites, and body condition.

Milk quality falling off in flavor: almost always chilling speed or equipment cleanliness. Chapters 8 and 11.

Udder and teat problems

Chapped or cracked teats. Common in cold, dry weather. Apply an udder balm after dipping. Painful teats cause kicking.

Warts. Usually viral and self-limiting. Do not pick at them.

Teat injuries. A stepped-on or torn teat is a veterinary matter, and quickly - teat lacerations heal poorly once the tissue retracts.

A blocked teat. Milk will not flow from one teat. May be a clot, a pea-sized obstruction, or scar tissue. Warm compresses and gentle massage first; a vet if it persists. Do not force anything into the teat canal - teat cannulas introduce infection and should only be used by someone trained.

Hard, uneven quarters. Suspect mastitis. Chapter 7.

Edema after freshening. Swelling in the udder in the days around birth is common, especially in first-fresheners. Massage, gentle exercise, and frequent milking. It resolves.

Behavior problems

Kicking. Rule out pain first - a sore teat, mastitis, or a chapped crack. Then consider technique, cold hands, flies, and inexperience. Hobbles while training; consistency and calm always.

Refusing the stand. Usually a bad experience. Rebuild with grain and short, calm sessions.

Holding milk. Stress or pain or a distraction. Also normal if she is saving it for a calf nearby.

Restlessness. Flies, an uncomfortable stand, an inconsistent routine, or something new in the environment.

The rule: an animal that suddenly behaves badly usually has a reason. Look for it before treating it as defiance.

When something is wrong with the milk

Clots or flakes: mastitis. Chapter 7.

Blood-tinged: often a broken capillary, common in a fresh animal and usually harmless if it clears in a few days. Persistent blood is a vet call.

Watery or thin: mastitis, or milk from a quarter that is drying up.

Off flavor:

- Slow chilling - the most common cause by far
- Dirty equipment
- Feed - onion, garlic, brassicas
- Buck exposure
- Mastitis
- Light exposure, which oxidizes milk fat and produces a distinctly cardboard-like flavor. Store milk in the dark
- Rancidity from excessive agitation, or from milk pumped through long hoses

Milk that will not culture or will not make cheese: chlorine residue on equipment, antibiotic residue in the milk, or milk that was frozen. Chapter 16.

PART FOUR - HANDLING THE MILK

Chapter 11: Straining, Chilling, and Storage

The twenty minutes after milking determine how the milk tastes and how long it keeps.

Strain immediately

Right after milking, before anything else.

Set a stainless strainer with a fresh disposable filter disc over a clean jar and pour the milk through.

What it removes: hair, dust, bits of bedding, and the occasional fly. Not bacteria - a filter does not sterilize anything.

One filter per straining. They are cheap. A reused filter is a bacterial culture.

A cloth is not adequate unless it is laundered and sanitized between every use, which is more work than buying filters.

Chill fast - the most important step

Milk leaves the animal at about 101 degrees F, which is close to a perfect incubation temperature. Bacterial counts double roughly every twenty to thirty minutes at that temperature.

Get it below 40 degrees F as fast as you can. Target: within thirty minutes of milking.

A refrigerator alone is too slow. A half-gallon jar of warm milk set in a refrigerator takes several hours to come down, and those hours are spent in the bacterial danger zone.

The ice bath method - do this:

1. Fill a sink, cooler, or tub with cold water and ice
2. Set the jar of strained milk in it, up to the shoulder
3. Stir the milk occasionally, or swirl the jar
4. Twenty to thirty minutes brings it near refrigerator temperature
5. Then move it to the refrigerator

A frozen water bottle dropped into the jar chills from the inside as well. Use a clean, sanitized bottle.

Coldest part of the refrigerator, at the back, away from the door.

This single practice separates milk that tastes clean and lasts ten days from milk that tastes off in three. More than breed, more than feed, more than anything else you control after the pail.

Storage

Glass, ideally. Nonreactive, doesn't hold odors, easy to sanitize, and you can see the cream line. Half-gallon wide-mouth canning jars are the standard.

Sealed. Milk absorbs odors from a refrigerator aggressively - onions, leftovers, anything.

Dark. Light oxidizes milk fat and produces a distinct cardboard or metallic off flavor within hours. This is why store milk moved from glass to opaque jugs. Keep jars in a closed refrigerator, not on a sunny counter.

Label with the date. You will have several jars going and it stops mattering fast.

Shelf life:

Storage

Milk	Refrigerated life
Raw, chilled fast, clean equipment	7-10 days, sometimes 2 weeks
Raw, chilled slowly	3-5 days
Pasteurized at home	10-14 days

Raw milk sours rather than rotting, because the lactic acid bacteria present outcompete putrefying organisms. Soured raw milk is not dangerous the way spoiled pasteurized milk is, and it is usable - it makes decent baking milk, and it can be cultured. Pasteurized milk that goes bad genuinely rots, because the protective flora were killed along with everything else.

That said: if it smells wrong to you, do not drink it.

Freezing milk

It works, with a caveat.

Method: freeze in plastic or in freezer-safe glass, leaving headspace for expansion. Milk expands substantially and will crack a full jar. Freeze in quantities you will use.

Life: 3 to 6 months.

Thaw in the refrigerator, slowly. Shake well - separation is normal.

The caveat: frozen and thawed milk does not make good cheese. Freezing damages the protein structure and the curd will be weak or will not form properly. It is fine for drinking, cooking, baking, and feeding animals.

Freezing is how you bridge the dry period. Build a stock in the last months of lactation.

Batches and mixing

Do not add warm milk to a jar of cold milk. It raises the temperature of the whole jar back into the danger zone.

Chill each milking separately, then combine once both are cold.

Keep track of which jar is oldest. Front to back rotation.

Reading milk quality

Good milk: clean, sweet, slightly nutty. A distinct cream line in cow milk after 12 to 24 hours. No off notes.

Common off flavors and their causes:

Reading milk quality

Flavor	Cause
Sour, acidic	Old, or chilled too slowly
Barny, unclean	Dirty equipment, dirty udder, dirty environment
Cardboard, metallic	Light exposure oxidizing the fat
Onion, garlic, bitter	Feed - check the pasture
Goaty, musky	Buck exposure - usually the housing, not the goat
Salty, off	Mastitis
Rancid	Excessive agitation, or long pumping through hoses
Chemical	Cleaning residue not rinsed

Almost every off flavor is a management problem with a specific fix. Milk is not inherently bad-tasting.

Chapter 12: Raw Milk, Honestly

The most contested topic in home dairying. Here are the facts, without a side.

The risk is real

Unpasteurized milk can carry pathogens, including Listeria monocytogenes, Salmonella, E. coli O157:H7, Campylobacter jejuni, Brucella, Coxiella burnetii (Q fever), and Mycobacterium bovis.

The CDC and FDA advise against consuming raw milk, and outbreaks are documented every year in the United States.

The consequences, when illness occurs, can be serious. E. coli O157:H7 causes hemolytic uremic syndrome, a cause of kidney failure in young children. Listeria in pregnancy causes miscarriage, stillbirth, and neonatal infection. Campylobacter and Salmonella cause severe gastroenteritis and occasionally lasting complications.

The highest-risk groups are: pregnant women, children under five, adults over sixty-five, and anyone immunocompromised.

Pasteurization was one of the great public health interventions of the twentieth century, and it was introduced because milk was killing people at scale. That history is real and it deserves respect.

What reduces the risk

Your own animal, well managed, is a meaningfully different proposition from bulk milk pooled from many herds. Not risk-free - meaningfully different.

What actually matters:

A healthy, tested animal. Brucellosis and tuberculosis are the historically serious ones. Ask your vet what testing is standard where you are, and do it.

A clean environment. Most contamination comes from manure. Dry bedding, clean flanks, a clean milking area.

Clean, dry teats before milking. Water running down a wet teat carries whatever was on the outside of the animal into the pail.

Stripping the first squirts away from the pail, since the teat canal holds the highest bacterial count.

Clean equipment, properly washed and air dried.

Fast chilling. Pathogens that get into milk multiply at body temperature and barely multiply at 38 degrees F.

No mastitis milk. An infected quarter is producing bacteria directly into the pail.

Single-animal, single-herd milk. Pooling multiplies exposure.

What none of that does is sterilize the milk. These practices reduce bacterial load substantially. They do not guarantee that a given jar is pathogen-free.

The honest summary

If you are a healthy adult drinking milk from your own tested animal, handled as described in Chapters 8 and 11, the risk is low. Many people do this for decades without incident.

It is not zero, and anyone telling you it is - in either direction - is selling something.

If someone in your household is pregnant, very young, elderly, or immunocompromised, pasteurize. The downside of pasteurizing is a slight change in flavor. The downside of a listeria infection in pregnancy is catastrophic. That is not a close call.

Decide deliberately, with the actual facts, for your actual household. Do not let either camp's confidence make the decision for you.

Pasteurizing at home

Straightforward, and it does not require special equipment.

Low-temperature (batch) pasteurization - preferred

145 degrees F, held for 30 minutes.

1. Pour milk into a stainless pot set inside a larger pot of water - a double boiler arrangement. Direct heat scorches
2. Heat slowly, stirring, until the milk reaches 145 degrees F
3. Hold at 145 degrees F for a full 30 minutes. Stir occasionally and watch the thermometer - you must maintain the temperature, not just touch it
4. Chill immediately in an ice bath, down below 40 degrees F as fast as possible
5. Store in clean sanitized jars

Why this method: gentler on milk proteins, so pasteurized milk still makes cheese. Flavor change is modest.

High-temperature short-time

161 degrees F for 15 seconds, then immediate chilling.

Faster, and harder to control at home without a proper setup. More protein denaturing, so it performs less well for cheese.

What does not work

UHT - ultra-pasteurized - milk cannot be used for cheese. The proteins are denatured so thoroughly that the curd will not form. This is why store-bought organic milk, which is often UHT, fails at mozzarella. Not relevant if you are pasteurizing your own gently, and very relevant if you ever buy milk to make cheese.

Boiling pasteurizes and ruins the flavor and the cheesemaking properties.

A microwave heats unevenly and cannot be relied upon.

After pasteurizing

Chill fast. Same rule as raw milk, and arguably more important - pasteurized milk has lost its protective flora, so recontamination has less competition.

Shelf life is somewhat longer than raw, 10 to 14 days.

Cultured products work fine. Yogurt, kefir, and cheese all work with home-pasteurized milk. Many cheesemakers prefer it because the results are more consistent.

Cheese and the 60-day rule

Under federal regulation, cheese made from raw milk may be sold in the US only if it has been aged at least 60 days at 35 degrees F or above. The aging period, combined with the salt and acidity of the cheese, reduces pathogen risk.

This applies to selling. For your own household, you make what you choose.

Note that the 60-day rule applies to aged cheeses. Fresh raw-milk cheeses - chevre, mozzarella, ricotta - cannot legally be sold in the US. Chapter 21.

What I would actually do

Since it would be evasive not to say:

Test the animal. Keep the milking clean and chill fast. Drink it raw if the household is healthy adults who want it.

Pasteurize for anyone pregnant, for small children, for elderly family members, and for anyone whose immune system is compromised - and pasteurize any milk you give to guests, because they did not get to inspect your barn.

That is a defensible position that takes the science seriously without pretending the practice is reckless.

Chapter 13: Cream and Separation

Where cow milk and goat milk part company.

Why cow milk separates

Milk fat is suspended as globules. In cow milk those globules are relatively large and they carry proteins that cause them to cluster and rise. Left still and cold, the fat floats.

In 12 to 24 hours in the refrigerator, a clear cream line forms in a jar of cow milk - often an inch or more in a jar of Jersey milk.

Why goat milk does not

Goat milk is naturally homogenized. The fat globules are smaller and lack the clustering proteins, so they stay suspended.

Some cream does rise in goat milk, very slowly and incompletely, over a day or two. It is thin, hard to skim cleanly, and there is not much of it.

For meaningful cream from goat milk, you need a mechanical cream separator. $200 to $600. It spins the milk and separates cream from skim by density.

If cream and butter matter to you, that is an argument for a cow - or for budgeting a separator.

Skimming cow cream

The ladle method. Let the milk sit 12 to 24 hours cold. Skim the cream off the top with a small ladle or a shallow spoon. Simple, and you inevitably take some milk with it.

The siphon method. Draw the skim milk out from the bottom with a siphon or a turkey baster, leaving the cream behind in the jar. Cleaner separation.

A cream separator jar - a jar with a spigot at the bottom - does the same thing conveniently. Drain skim milk from the bottom until cream reaches the tap.

A wide shallow pan gives more surface area and a faster, more complete rise than a tall jar. Traditional, and it takes refrigerator space.

Do not shake the jar before skimming, obviously. And be gentle carrying it.

Mechanical separators

How they work: a spinning bowl uses centrifugal force to throw the denser skim milk outward and channel the lighter cream inward, with two spouts.

Advantages:

- Works on goat milk
- Far more complete extraction
- Immediate - no waiting for cream to rise
- Lets you control the fat content of the cream

Disadvantages:

- $200 to $600
- Many parts to clean, every time. This is the real cost and it is why separators end up in barn corners
- Electric or hand crank

Worth it if: you have goats and want butter, or you have a cow and want maximum cream yield, or you make a lot of butter.

Not worth it if: you have a cow and skimming by ladle covers your needs. Most family cow owners never buy one.

Cream fat content

Skimmed cow cream typically runs 18 to 30 percent fat depending on how carefully you skim.

Separator cream can be adjusted, commonly 35 to 45 percent.

For whipping, you want roughly 30 percent or higher. Hand-skimmed Jersey cream usually whips.

For butter, higher fat means faster churning and better yield. Below about 30 percent, churning is slow and frustrating.

What to do with skim

Do not throw it out. Skim milk is still most of the protein and all of the lactose.

- Drinking, if you like skim
- Yogurt - makes a firm, tangy yogurt
- Most cheeses. Many traditional farmhouse cheeses were made from skimmed milk with the cream sold separately
- Baking
- Feeding to pigs, chickens, or dogs - an excellent protein supplement
- Ricotta and paneer

Storing cream

Refrigerated, sealed, in glass. A week or more.

Cream ages well for butter. Cream that is a few days old actually churns more readily than very fresh cream, and it develops flavor.

Freezing cream is possible and it changes the texture. Fine for cooking, less good for whipping.

Collect cream over several days in a single jar in the refrigerator until you have enough to churn. This is the normal rhythm - skim daily, churn weekly.

PART FIVE - IN THE KITCHEN

Chapter 14: Butter and Buttermilk

The first thing most people make, and the most satisfying.

The principle

Cream is fat globules suspended in liquid. Agitate it and the globules collide, the membranes rupture, and the fat coalesces into a mass. What is left behind is buttermilk.

That is all churning is: breaking the emulsion.

Temperature is the whole game

55 to 60 degrees F is the target. This is the single most common reason butter fails to come.

Too cold, below about 50 degrees F: the fat is too hard, and you will churn for an hour and produce nothing but very stiff whipped cream.

Too warm, above about 65 degrees F: the fat is too soft. The butter comes greasy, soft, and hard to work, and yield suffers.

Cream straight from the refrigerator is too cold. Set it on the counter for an hour or two and check it with a thermometer. This one adjustment fixes most butter problems.

Sweet cream or cultured

Sweet cream butter: churn the cream as is. Mild, clean, familiar. This is American-style butter.

Cultured butter: culture the cream first - add a couple of tablespoons of cultured buttermilk, kefir, or a mesophilic culture per quart, and leave it at room temperature 12 to 24 hours until it thickens and smells tangy. Then chill to

55-60 degrees F and churn.

Cultured butter is European-style. Deeper, more complex, slightly tangy. It also churns faster, because the acidity helps break the emulsion. Worth doing - the flavor difference is substantial and the extra work is essentially nothing.

Churning

Stand mixer. Whisk attachment, medium speed, splash guard on. The most practical method for most kitchens. Ten to twenty minutes.

Food processor. Fastest - often under five minutes. Loud, and it goes from cream to butter suddenly.

A jar. Fill a quart jar halfway, seal, and shake. Fifteen to thirty minutes and it is genuinely tiring. Excellent for children, who find it delightful for the first ten minutes.

A butter churn, hand crank or electric. If you make butter weekly, worth having.

Blender. Works, and tends to overheat the cream.

The stages

You will pass through all of these:

1. Liquid cream
2. Soft peaks - whipped cream
3. Stiff peaks
4. Grainy, over-whipped, starting to look curdled - this is the point where people think they have ruined it. Keep going
5. The break - suddenly, and it really is sudden, the butter separates into yellow clumps swimming in thin white buttermilk. You will hear the sound change

Washing the butter

This step is not optional and it is the one people skip.

Why: residual buttermilk trapped in the butter goes rancid quickly. Unwashed butter turns within days. Properly washed butter keeps for weeks.

How:

1. Drain off the buttermilk through a strainer. Save it
2. Add ice-cold water to the butter and work it - press and fold with a spatula, a wooden paddle, or clean hands
3. Pour off the cloudy water
4. Repeat until the water runs clear. Usually three to five rinses
5. Work out the remaining water by pressing and folding, tilting to drain. The more water you remove, the longer it keeps

Use ice-cold water. Warm water melts the butter and makes it greasy.

Salting and shaping

Salt to taste, roughly 1 to 2 percent by weight - about a quarter teaspoon per half pound as a starting point. Use fine non-iodized salt and work it in evenly.

Salt adds shelf life as well as flavor. Unsalted butter keeps less well and freezes better.

Shape into a log in parchment, press into a mold or a butter bell, or just pack into a container.

Yield

Roughly 2 cups of cream yields about 8 ounces of butter plus about 1 cup of buttermilk, though it varies substantially with fat content. Higher-fat cream yields proportionally more.

A gallon of Jersey milk yields perhaps a quart of cream, which yields perhaps a pound of butter.

Storage

- Refrigerated: 2 to 3 weeks salted, less unsalted
- Frozen: 6 to 12 months, well wrapped
- On the counter in a butter bell - a crock where the butter sits inverted in a water seal - for a week or so in a cool kitchen. Salted butter only

Buttermilk

True buttermilk is what is left after churning. Thin, slightly tangy if the cream was cultured, and quite different from the thick cultured buttermilk sold in stores.

Uses: biscuits, pancakes, cornbread, soda bread, fried chicken brine, salad dressing, marinades, or drinking.

From cultured cream it is tangier and better. From sweet cream it is mild and closer to skim milk.

Keeps about a week refrigerated. Freezes acceptably.

Ghee

Butter with the water and milk solids removed. Shelf stable at room temperature, high smoke point, nutty flavor.

1. Melt butter in a heavy pan over low heat
2. It foams as water evaporates
3. Milk solids sink and begin to brown - this is where the flavor comes from
4. When the bubbling quiets and the solids on the bottom are golden brown, remove from heat

5. Strain through cheesecloth into a clean jar

6. Do not let the solids burn, which turns it bitter

Keeps months at room temperature, longer refrigerated. A good use for butter you have too much of.

Troubleshooting

It will not break. Cream too cold, cream too low in fat, or not enough time. Warm it toward 60 degrees F and keep going.

Greasy, soft butter. Cream too warm. Chill everything and rework it in ice water.

Low yield. Low-fat cream, or butter left behind in the buttermilk. Some loss is normal.

Goes rancid quickly. Not washed enough. More rinses next time.

Won't come at all from goat cream. Goat cream churns differently and can be stubborn. Culture it first, get the temperature right, and be patient.

Chapter 15: Yogurt, Kefir, and Cultured Cream

The easiest transformations, and the ones you will do weekly.

Book 20 in this series covers fermentation principles in depth. This chapter is the dairy application.

Yogurt

Ingredients: milk, and 2 tablespoons of live-culture plain yogurt per gallon, or a powdered starter.

Method:

1. Heat the milk to 180 degrees F, stirring often. This denatures the whey proteins so they set into a firm gel instead of staying liquid. Do not skip this - it is the difference between yogurt and thin cultured milk. Watch it; milk scorches and boils over fast
2. Cool to 110-115 degrees F. An ice bath speeds this considerably. Above 120 degrees F kills the culture; below 100 degrees F it works too slowly
3. Temper the starter - whisk the 2 tablespoons of yogurt with a cup of the warm milk, then stir that back in. Prevents clumping
4. Incubate at 105-110 degrees F for 4 to 12 hours
5. Refrigerate at least 4 hours before eating. It firms considerably as it chills
6. Save 2 tablespoons for the next batch

Ways to hold 110 degrees F:

- An oven with only the light on - verify yours actually holds near 110 degrees F

- A cooler filled with 110 degrees F water, jars sealed and submerged to the neck
- An insulated bag, or a pot wrapped in towels somewhere warm
- A yogurt maker or an appliance with a yogurt setting
- A heating pad on low under a towel-wrapped pot

Timing and flavor:

- 4 to 6 hours: mild, creamy
- 8 to 12 hours: tangy, firmer
- Longer: sour, and it may separate

Goat milk yogurt is naturally thinner, because goat milk protein sets more softly. Strain it, or add a few tablespoons of powdered milk before heating, or accept a drinkable yogurt - which is what much of the world eats anyway.

Greek yogurt: strain finished yogurt through butter muslin over a bowl. Two to four hours for Greek thickness, overnight for a spreadable cheese. Save the whey.

Carrying the culture forward: each generation loses a little vigor. Every ten to twenty batches, or whenever it gets slow and thin, restart from fresh commercial yogurt or a packet.

Kefir

Thinner and tangier than yogurt, with a much broader range of organisms, and considerably easier.

You need milk kefir grains - small springy clusters that look like cauliflower. They cannot be made from scratch. Get them from another homesteader or buy them.

Method:

1. Put 1 to 2 tablespoons of grains in a jar

2. Add 2 to 4 cups of milk

3. Cover loosely

4. Leave at room temperature 12 to 48 hours. Twenty-four at 70 degrees F is typical

5. Done when thickened and tangy. Separation into curds and whey means it went long - still fine, just stir

6. Strain through a plastic or nylon strainer. Avoid prolonged contact with reactive metal

7. Put the grains straight into fresh milk and repeat

No heating, no temperature control, no thermometer. This is why kefir suits a home dairy so well - it absorbs surplus milk with essentially no effort.

The grains multiply. Share them, eat them, or scale up.

Taking a break: grains in fresh milk in the refrigerator hold two to three weeks.

Uses: drinking, smoothies, baking in place of buttermilk, or strained through cloth into a soft spreadable cheese.

Sour cream

2 tablespoons of cultured buttermilk per 2 cups of cream. Stir, cover, leave at room temperature 12 to 24 hours until thickened. Refrigerate.

Or use a mesophilic cheese culture per its directions.

Creme fraîche

The same thing, essentially, and better.

2 tablespoons cultured buttermilk in 2 cups heavy cream. Room temperature 12 to 24 hours. Refrigerate.

Why it beats sour cream: it does not curdle when heated. You can stir it into a hot pan sauce or a soup and it stays smooth. Anyone who cooks will use this constantly.

Cultured butter

Chapter 14. Culture the cream first, then churn.

Labneh

Salted strained yogurt. Stir a teaspoon of salt into a quart of yogurt, tie it in butter muslin, and hang it over a bowl for 12 to 24 hours.

The result is a thick spreadable cheese. Roll it into balls, cover with olive oil, and store in the refrigerator for weeks. Excellent with bread and good oil.

Troubleshooting

Yogurt is thin. Did not heat to 180 degrees F, incubated too cool or too briefly, weak starter, or goat milk. Strain it and call it Greek.

Yogurt is grainy. Heated too hot, or the starter went in above 120 degrees F.

Yogurt never set. Culture killed by heat, incubation temperature dropped, or antibiotic residue in the milk. Verify temperatures with a thermometer rather than trusting an appliance.

Kefir will not thicken. Too few grains for the volume, too cold, or weak grains. More grains, less milk, warmer room.

Kefir tastes yeasty. Yeast outrunning bacteria. Rinse the grains gently in filtered water, use fresh milk, ferment cooler.

Anything ropey or slimy. Occasionally a stressed culture. Start fresh.

Everything fails at once. Suspect chlorine on your equipment, or antibiotic residue in the milk. Both stop cultures cold.

Chapter 16: Cheese Without Culture

The fastest cheeses there are. Milk, heat, acid, thirty minutes.

How acid-set cheese works

Heat milk, add an acid, and the casein proteins destabilize and clump into curds, leaving whey behind. No culture, no rennet, no aging, no waiting.

These cheeses do not melt - the protein structure is set by acid rather than rennet - which makes them ideal for frying and for holding shape in a dish.

Whole milk ricotta

Ingredients: 1 gallon milk - 1/4 cup lemon juice or white vinegar - 1 teaspoon salt

Method:

1. Heat the milk slowly to 185-195 degrees F, stirring occasionally so it does not scorch. Use a heavy pot or a double boiler
2. Remove from heat. Stir in the acid gently, once or twice. Do not stir vigorously - you want the curds to form in large soft masses
3. Let it sit undisturbed 10 minutes. Curds separate and the whey turns yellowish-green and translucent
4. Ladle the curds into a colander lined with butter muslin
5. Drain 15 to 60 minutes. Less for creamy, more for firm
6. Salt to taste. Stir gently

Yield: roughly 1 1/2 to 2 cups from a gallon.

Uses: lasagna, stuffed pasta, on toast with honey, in pancakes, sweetened as a dessert.

Keeps about a week refrigerated.

Note on the name: true traditional ricotta is made from the whey left over from another cheese - ricotta means recooked. This whole-milk version is what most people make at home and it is richer. The whey version is in Chapter 20.

Paneer and queso blanco

Firmer, pressed, and it does not melt - which is the point.

Ingredients: 1 gallon milk - 1/4 to □ cup lemon juice or vinegar - 1 teaspoon salt

Method:

1. Heat milk to 185-190 degrees F
2. Remove from heat, stir in acid, let sit 10 minutes
3. Pour into a colander lined with butter muslin
4. Rinse the curds briefly with cool water to remove the acid flavor
5. Gather the cloth, twist to squeeze out whey
6. Press. Wrap in the cloth, shape into a flat block, and set a weight on it - a pot of water works - for 1 to 3 hours
7. Refrigerate

Yield: roughly 1 pound from a gallon.

Uses: cube and fry until golden, then add to curries. Grill it. Crumble over salads. It holds its shape at any temperature, which nothing else in this book does.

Keeps about a week.

Queso fresco

Same as paneer, pressed more lightly, salted more heavily. Crumbly and mild. Excellent on tacos, beans, and grilled corn.

Cottage cheese

A simple acid-set version:

1. Heat 1 gallon milk to 190 degrees F
2. Add 1/4 cup vinegar, stir gently, rest 10 minutes
3. Drain in a colander
4. Rinse thoroughly with cold water, which both cools the curds and stops the acid
5. Break the curds up with your fingers
6. Salt, and stir in a few tablespoons of cream or milk to dress it

Traditional cottage cheese uses a mesophilic culture and takes overnight, and it has better flavor. This version takes thirty minutes.

Mascarpone

Cream, acid, and nothing else.

1. Heat 2 cups heavy cream to 185 degrees F in a double boiler
2. Add 1 tablespoon lemon juice. Stir gently
3. Hold at 185 degrees F for 5 minutes, stirring occasionally. It will thicken slightly
4. Cool, then refrigerate in a cloth-lined strainer for 12 to 24 hours

Rich, thick, and slightly sweet. Tiramisu, desserts, or spread on anything.

Acid choices

Lemon juice - clean, slight citrus note. The default.

White vinegar - sharper, most neutral once rinsed. Cheapest.

Cider vinegar - adds a distinct flavor. Fine in some applications.

Citric acid - most neutral of all, dissolved in a little water first. Sold with cheesemaking supplies.

Cultured whey or buttermilk - traditional, slower, more complex flavor.

Troubleshooting

No curds formed.

- Ultra-pasteurized (UHT) milk. The most common cause. The proteins are denatured and will not curd. Check your carton. This is why store-bought organic milk often fails
- Not hot enough - verify with a thermometer
- Not enough acid - add a little more, a tablespoon at a time
- Milk was previously frozen

Curds are tiny and grainy. Stirred too vigorously, or too much acid. Stir gently and use the minimum acid that works.

Cheese tastes sour. Too much acid, or not rinsed. Rinse the curds.

Yield is low. Normal - cheese is a small fraction of milk. A gallon giving a pound is about right. If it is much less, check the temperature and acid.

Rubbery texture. Overheated, or pressed too hard.

Why start here

These cheeses are nearly foolproof, they take half an hour, and they use up surplus milk immediately. When you are suddenly getting four gallons a day and the refrigerator is full, a batch of paneer is the fastest way to convert milk into something that keeps and something you will actually eat.

Chapter 17: Chevre and Cream Cheese

The first cultured cheeses. Almost as easy as Chapter 16, and considerably better.

What changes

These use a mesophilic culture - bacteria that work at room temperature - and a tiny amount of rennet. The culture acidifies slowly over many hours, producing far more flavor than a quick acid set, and the rennet firms the curd.

They are extremely forgiving. Chevre is the cheese to start with if you have never used a culture.

Supplies

Mesophilic culture. Direct-set packets, sold by cheesemaking suppliers. Store in the freezer. A packet typically does 1 to 2 gallons. Some suppliers sell a dedicated chevre culture with the rennet already included, which is even simpler.

Rennet. Liquid or tablet; animal, microbial, or vegetable - any works here. Store cold. It loses potency over time, so replace it annually.

Butter muslin. Fine-weave cheesecloth. Grocery-store cheesecloth is too open and your curd will go through it. Butter muslin is worth ordering.

Non-chlorinated water for diluting rennet. Chlorine kills cultures.

A non-reactive pot - stainless or enamel. Not aluminum.

Chevre

Traditionally goat cheese, and it works perfectly well with cow milk, where it is called fromage blanc.

Ingredients: 1 gallon milk - 1/8 teaspoon mesophilic culture - 1 drop liquid rennet diluted in 1/4 cup cool non-chlorinated water - salt

Method:

1. Warm the milk to 86 degrees F. Gently. Remove from heat
2. Sprinkle the culture over the surface. Let it rehydrate 2 minutes, then stir gently up and down for 20 seconds
3. Add the diluted rennet. Stir gently for 20 seconds. The amount is genuinely tiny - one drop per gallon. More rennet gives a rubbery curd
4. Cover and leave at room temperature, 68-72 degrees F, for 12 to 18 hours
5. It is ready when it looks like thick yogurt, has pulled slightly from the pot sides, and a little clear whey has pooled on top
6. Ladle into a butter-muslin-lined colander. Or tie the cloth into a bag and hang it
7. Drain 6 to 12 hours at room temperature or in the refrigerator. Less for soft and spreadable, more for firm and crumbly
8. Salt to taste - roughly 1 teaspoon per pound - and mix

Yield: about 1 1/2 pounds from a gallon.

Keeps 1 to 2 weeks refrigerated.

Flavorings, mixed in after salting: cracked pepper, fresh herbs, garlic, chives, lemon zest, honey, dried cranberries. Roll logs in herbs or pepper.

This is the highest-return cheese in the book. Fresh chevre made this morning is a genuinely different food from anything in a store, and the process is nearly automatic.

Cream cheese

Ingredients: 1 quart whole milk + 1 quart heavy cream (or 2 quarts whole milk for a lighter version) - 1/8 teaspoon mesophilic culture - 1 drop rennet in 1/4 cup cool water - salt

Method: identical to chevre.

1. Warm to 86 degrees F
2. Culture, rest 2 minutes, stir
3. Diluted rennet, stir
4. Cover, 12 to 18 hours at room temperature
5. Drain in butter muslin 6 to 12 hours - longer for firmer
6. Salt, and beat smooth with a spoon or a mixer for a silky texture

Richer than chevre, and it behaves like store cream cheese in most applications.

Fromage blanc

Chevre made with cow milk. Same process. Milder and slightly sweeter.

Neufchâtel

Cream cheese made with milk only, no added cream. Lighter, tangier.

Making it firmer or softer

Softer and more spreadable: drain less time. Refrigerate during draining to slow it.

Firmer and crumbly: drain longer, up to 24 hours. Press lightly with a small weight.

For a log you can slice: drain to firm, salt, roll tightly in plastic wrap, and refrigerate several hours.

Storing

Refrigerated in a sealed container: 1 to 2 weeks.

Under oil: roll into balls, cover completely with olive oil, and refrigerate. Keeps a month or more and the oil takes on the flavor. Keep it refrigerated - cheese under oil at room temperature is an anaerobic low-acid environment and not a risk worth taking.

Freezing works and changes the texture - it becomes crumbly. Fine for cooking, less good for spreading.

Troubleshooting

Never set after 18 hours. Room too cold - below about 65 degrees F it stalls. Move somewhere warmer and give it more time. Or the culture was old or dead, or there was chlorine on the equipment, or antibiotic residue in the milk.

Set too fast and separated hard. Room too warm, or too much rennet.

Rubbery texture. Too much rennet. One drop per gallon, diluted.

Bitter. Too much rennet, or over-ripened.

Curd went through the cloth. Cheesecloth too open. Use butter muslin.

Slimy or off-smelling. Contamination. Discard, sanitize everything, and check your milk handling.

Chapter 18: Mozzarella

The showpiece, and the one that fails most often for beginners. Here is why it fails and how not to.

The milk problem, first

Mozzarella will not work with ultra-pasteurized (UHT) milk. The proteins are denatured beyond the point where they can form a stretchable curd. You will get a soupy mess and conclude you did something wrong.

Use raw milk, or milk you pasteurized gently at 145 degrees F. If you buy milk, check the carton - much organic milk is UHT and most cream is.

This single fact accounts for most failed mozzarella.

The 30-minute method

Ingredients:

- 1 gallon whole milk, not UHT
- 1 1/2 teaspoons citric acid dissolved in 1/4 cup cool non-chlorinated water
- 1/4 teaspoon liquid rennet diluted in 1/4 cup cool non-chlorinated water
- 1 teaspoon salt

Method:

1. Pour the citric acid solution into cold milk. Stir well. Doing this while the milk is cold is important - added to warm milk, the acid causes uneven curdling
2. Heat slowly to 90 degrees F, stirring gently. The milk may look slightly curdled. Normal
3. Remove from heat. Add the diluted rennet. Stir gently up and down for 30 seconds, then stop completely

4. Cover and rest 5 to 10 minutes, undisturbed. Do not stir, do not bump the pot

5. Check for a clean break. Slip a knife into the curd at an angle and lift. It should split cleanly, like set gelatin, with clear whey filling the cut. If it is still soft and milky, wait another 5 minutes

6. Cut the curd into roughly 1-inch cubes with a long knife, cutting all the way to the bottom in a grid

7. Heat gently to 105 degrees F, stirring slowly and carefully. The curds firm and shrink and the whey turns yellowish

8. Drain. Ladle the curds into a colander and let the whey run off. Save the whey

9. Heat the curd to stretching temperature. Two ways:

- Hot water method (better control): heat whey or water to 175 degrees F. Submerge the curd in it, in a bowl, for a few minutes. Work it with a spoon
- Microwave method (faster): microwave the curd 1 minute, drain the whey, fold it over on itself, microwave 30 seconds more, drain and fold again

10. The curd is ready to stretch when its internal temperature is about 135 degrees F. It will look glossy and begin to pull like taffy

11. Stretch and fold repeatedly - pull it out, fold it back, pull again - until smooth and shiny. Do not overwork it, or it turns tough

12. Salt during stretching, or brine the finished ball

13. Shape into a ball and drop into ice water for 15 minutes to set

Yield: about 1 pound from a gallon.

Eat it the same day. Fresh mozzarella is dramatically better within hours of making it, and it declines noticeably after a day or two.

Storing

In lightly salted whey or brine, refrigerated, 2 to 3 days.

Freezing works if you intend to cook with it - shredded and frozen mozzarella melts fine on pizza. It ruins the texture for eating fresh.

Burrata, if you want to show off

Make mozzarella, but before shaping, tear some of the fresh curd into shreds and mix with heavy cream. Then form a thin mozzarella pouch, fill it with the cream-and-curd mixture, and seal the top.

Genuinely impressive and not much harder.

Troubleshooting

Curd never set / no clean break.

- UHT milk. The most likely cause
- Old or weak rennet - replace annually
- Chlorinated water used to dilute the rennet
- Not enough rennet
- Milk was previously frozen

Curd is soft and will not hold together. Not enough acid, or not hot enough at the stretching stage.

Will not stretch. Almost always temperature - the curd must reach about 135 degrees F internally. Check with a thermometer rather than guessing. Also possible: too much or too little acid.

Stretches then breaks and crumbles. Too acidic, or overworked.

Tough and rubbery. Overworked during stretching, or heated too hot. Stretch just until smooth and stop.

Tastes bland. Not enough salt. Salt during stretching, or brine it.

Grainy. Overheated, or the acid was added to warm milk.

Why it is worth the trouble

Mozzarella is the cheese where the home version most obviously beats the store version. Fresh mozzarella still slightly warm, torn onto a tomato with olive oil and salt, is one of the genuine arguments for keeping a dairy animal.

It also takes half an hour, which makes it the fastest way to turn a gallon of surplus milk into something people will fight over.

Chapter 19: Your First Aged Cheese

Where home dairying gets serious. Be honest with yourself about whether you want this.

What aging requires

Space with controlled conditions. Aged cheese needs roughly 50 to 55 degrees F at 80 to 90 percent humidity for weeks to months. A normal refrigerator is too cold and far too dry.

Options:

- A dedicated mini fridge with an external temperature controller. The standard solution, and it costs perhaps $150 to $250 total. A shallow pan of water inside handles humidity
- A root cellar or unheated basement, if it holds the right range. Book 18 in this series covers cellar conditions
- A cave, if you are fortunate

A press. For hard cheeses. Commercial presses run $80 to $300; a serviceable one can be built from two boards, threaded rod, and a spring, or improvised with a weighted pot.

Time and attention. A wheel needs turning, and often wiping or brushing, over weeks and months.

Patience. Your first cheese will be edible in two months and better at six. You will not know whether you succeeded for a long time.

If that sounds like a lot, it is. Fresh cheeses in Chapters 16 through 18 give you 90 percent of the daily value for 10 percent of the effort. Aged cheese is a distinct hobby that happens to use the same milk.

Additional supplies

- Mesophilic culture for most hard cheeses; thermophilic for Italian styles
- Rennet, fresh
- Cheese salt - non-iodized, no anti-caking agents
- Cheese molds or forms, sized to your batch
- Cheese cloths for pressing
- A curd knife - any long blade that reaches the pot bottom
- A thermometer
- Cheese wax, or bandaging cloth and lard, or nothing if you go natural rind

A basic farmhouse cheese

This is a simplified pressed cheese in the general style of a farmhouse cheddar. It is a real recipe and it is also a teaching recipe - the point is to walk the steps once.

Ingredients: 2 gallons whole milk - 1/2 teaspoon mesophilic culture - 1/2 teaspoon liquid rennet in 1/4 cup cool non-chlorinated water - 2 tablespoons cheese salt

1. Ripen. Warm milk to 86 degrees F. Add culture, rest 2 minutes, stir gently. Hold at 86 degrees F for 45 minutes. The culture is acidifying the milk.

2. Set. Add diluted rennet, stir gently 30 seconds, stop. Cover and hold at 86 degrees F for 45 minutes until you get a clean break.

3. Cut. Cut the curd into 1/2-inch cubes. Let rest 5 minutes.

4. Cook. Raise the temperature slowly to 102 degrees F over about 30 minutes - roughly 2 degrees every 5 minutes - stirring gently and constantly to keep curds from matting. Hold at 102 degrees F for 30 minutes, still stirring occasionally. The curds shrink and firm.

5. Drain. Pour off the whey through a cloth-lined colander. Savc the whey.

6. Cheddar, or not. For a cheddar-style texture: let the curd mass knit for 15 minutes, cut it into slabs, and stack and turn them every 15 minutes for about an hour, draining whey as it releases. This step develops the characteristic texture. For a simpler cheese, skip it.

7. Mill. Break the curd into roughly walnut-sized pieces.

8. Salt. Sprinkle the cheese salt over the curds and mix thoroughly. Salt draws out more whey, controls the culture, and preserves.

9. Press.

- 30 minutes at light pressure - roughly 10 pounds
- Remove, unwrap, flip, rewrap
- 1 hour at medium pressure - roughly 20 pounds
- Flip again
- 12 hours at full pressure - roughly 40 to 50 pounds

Flipping matters. It equalizes pressure and moisture.

10. Air dry. Remove from the press, unwrap, and set on a mat at room temperature for 1 to 3 days, turning daily, until the surface is dry to the touch.

11. Finish the rind. Three approaches:

- Wax: melt cheese wax and brush on two or three coats. Simplest, and it prevents mold entirely

- Bandage: wrap in cloth rubbed with lard or butter. Traditional, develops more complex flavor, allows some mold
- Natural rind: leave bare and brush off mold as it appears. Most work, most character

12. Age at 50-55 degrees F, 80-90 percent humidity.

- Turn the wheel every few days at first, then weekly
- Wipe or brush off unwanted mold
- 2 months minimum. 3 to 6 for a decent cheese. 6 to 12 for a good one

Yield: roughly 2 pounds from 2 gallons.

About mold on aging cheese

Some mold is normal and expected on a natural rind.

White and gray molds are generally harmless. Brush them off, or wipe with a cloth dampened in salt brine or vinegar.

Blue and green molds are usually harmless on a hard rind. Wipe them off.

Black, pink, or fuzzy red - remove that cheese and evaluate carefully. Pink and red growths in particular are worth taking seriously.

On a hard, low-moisture cheese, surface mold does not penetrate deeply, and cutting well below the affected area is standard practice. On a soft cheese, mold penetrates and the cheese is done.

Cracks in the rind let mold inside where you cannot manage it. Wax or bandage if your humidity is unreliable.

The 60-day rule

Raw milk cheese sold in the United States must be aged at least 60 days at 35 degrees F or above. The combination of time, salt, and acidity reduces pathogen risk.

This governs selling, not your own consumption. Chapter 21.

It also means fresh raw-milk cheeses - chevre, mozzarella, ricotta - cannot legally be sold in the US, which is why nearly all fresh cheese in American stores is pasteurized.

Going further

This chapter gets you through one wheel. Aged cheesemaking is genuinely deep, and if it catches you, get a dedicated book.

Worth owning: Home Cheese Making by Ricki Carroll, the standard beginner reference with a wide recipe range. Mastering Artisan Cheesemaking by Gianaclis Caldwell, for the serious next step - she explains why things happen rather than just what to do. The Art of Natural Cheesemaking by David Asher, for a traditional, culture-from-scratch approach.

Also join or find a cheesemaking forum. Diagnosing a wheel from a description is a skill, and experienced makers are generous with it.

Chapter 20: Whey, and Wasting Nothing

Every gallon of milk that becomes a pound of cheese leaves behind most of a gallon of whey. Here is what to do with it.

Two kinds of whey

Sweet whey comes from rennet-set cheeses - mozzarella, farmhouse cheddar, most hard cheeses. Around pH 6, mild and slightly sweet. The useful one.

Acid whey comes from acid-set and cultured products - yogurt straining, chevre, ricotta, cream cheese. Around pH 4.5, distinctly sour. Harder to use, and a genuine disposal question at volume.

Know which you have, because the uses differ.

Uses for sweet whey

Ricotta, properly. Traditional ricotta is made from whey - ricotta means recooked. Heat sweet whey to 195 degrees F, add a splash of vinegar, and small delicate curds form. Strain through butter muslin. The yield is small - perhaps a cup from a gallon of whey - and it is exceptional. This only works with sweet whey, and it must be fresh.

Baking. Replace water in bread, pancakes, biscuits, and pizza dough. It adds protein, improves browning, and slightly softens the crumb. One of the best uses there is.

Cooking liquid. Cook rice, pasta, oatmeal, grits, or beans in it.

Soup and stock base.

Soaking grains before cooking, which improves digestibility.

Smoothies.

Lacto-fermentation starter. A splash of whey can be used to start vegetable ferments. Book 20 covers fermentation - note that vegetable ferments do not require it and work fine on wild bacteria alone.

Uses for acid whey

More limited, and still useful.

Baking, where its acidity works with baking soda - it behaves much like buttermilk.

Marinades. The acidity tenderizes.

Animal feed. See below.

Plant watering, well diluted.

Feeding whey to animals

The best high-volume use, and the traditional one.

Pigs. The classic pairing. Pigs thrive on whey and historically dairy and pig farming went together for exactly this reason. Introduce gradually.

Chickens. Excellent protein supplement. Offer in a dish; do not let it sour in the sun.

Dogs. Fine in moderation.

Calves and kids. Sweet whey supplements well.

Introduce gradually in every case - a sudden large volume causes digestive upset.

Whey on the garden

Diluted, roughly 1 part whey to 4 or more parts water, it adds nitrogen, potassium, and calcium.

Do not pour undiluted whey on plants. The acidity and the sheer nutrient load can burn roots and disrupt soil biology.

Acidic whey suits acid-loving plants - blueberries, azaleas, tomatoes - better than it suits most vegetables.

Compost is usually better. Add whey to a compost pile with plenty of carbon material. Book 5 in this series covers composting.

Disposal, and why it matters

Do not dump large volumes of whey down a drain or on the ground.

Whey has a very high biological oxygen demand. In a septic system it overwhelms the bacteria. In a stream it strips oxygen and kills fish. On a small patch of ground it goes anaerobic, smells terrible, and kills the vegetation.

At small household volumes - a gallon or two a week - spreading it broadly on compost or pasture is fine.

At larger volumes, feed it to animals. That is what the animals are for.

The rest of the byproducts

Skim milk. Chapter 13 - yogurt, cheese, baking, animals.

Buttermilk. Chapter 14 - baking, drinking, culturing.

Excess milk generally. Freeze it, make hard cheese that keeps, feed it to livestock, give it away, or plan a smaller herd next time.

Manure. The genuinely valuable byproduct. Composted goat and cow manure is excellent garden fertilizer, and goat manure in particular is relatively mild and can be used with less aging than most. Book 17 covers soil and fertility.

The closed loop

This is the argument for a home dairy that has nothing to do with milk.

Milk feeds the family. Whey feeds the pigs and chickens. Pigs and chickens feed the family. Manure feeds the garden. The garden feeds the family and the animals. Pasture feeds the animals and the pasture is improved by their grazing and their manure.

None of this happens if you buy milk in a jug. It is not the reason to keep a dairy animal, and it is a real part of why the ones who do it keep doing it.

PART SIX - THE LONG GAME

Chapter 21: The Legal Side

Short chapter, and the one that will save you trouble.

Milk for your own household

Producing and drinking raw milk from your own animal is legal throughout the United States. No state prohibits you from drinking milk from your own cow or goat.

Everything that follows is about selling and sharing.

Selling raw milk

This varies enormously by state, and it changes. The landscape includes states that permit retail sale in stores, states that permit on-farm sales only, states that permit sales only for animal consumption, states that permit herd shares, and states that prohibit all sale.

Find out for your state before you sell a single jar. Search your state's name plus "raw milk law" and check with your state department of agriculture. Do not rely on what a neighbor tells you, on an internet forum, or on this book - these rules change and the penalties for getting it wrong can include fines and losing your animals.

Herd shares are an arrangement in which a buyer purchases a share of the animal and pays a boarding fee, receiving milk as an owner rather than a customer. They are explicitly legal in some states, explicitly illegal in others, and in an unresolved gray zone in several. Get local legal advice if you go this route.

Even where legal, selling usually brings requirements: licensing, inspection, testing, labeling, bottling standards, and record keeping.

Selling cheese

Raw milk cheese must be aged 60 days at 35 degrees F or above to be sold in the US. Federal rule.

Which means fresh raw-milk cheese cannot legally be sold - no raw chevre, no raw mozzarella, no raw ricotta.

Pasteurized cheese is subject to state food processing rules - which commonly means a licensed and inspected facility, not your kitchen.

Cottage food laws vary by state and most exclude dairy entirely because it is a higher-risk category. Do not assume a cottage food exemption covers cheese; check specifically.

Selling other products

Butter, yogurt, and kefir are generally treated as dairy products with the same processing requirements as cheese.

Eggs and produce are usually much easier. Book 13 in this series covers selling from the homestead.

Zoning and local rules

Check before you buy an animal:

- Is livestock permitted on your parcel?
- Are there animal unit limits based on acreage?
- Setback requirements for barns and manure storage?
- Are dairy animals treated differently from other livestock?
- Homeowners association covenants?
- Are there restrictions on roosters, bucks, or bulls specifically? Noise and odor ordinances often catch these
- Manure management requirements?

Municipal and county rules are where people get caught, usually after a neighbor complains.

Animal identification and testing

Some states require registration, premises identification, or specific testing for certain species. Interstate movement of livestock generally requires a health certificate and testing.

Ask your veterinarian and your state department of agriculture.

Liability

If you give milk away, you carry some risk. Someone who gets sick after drinking milk you provided may have a claim, regardless of whether money changed hands.

Practical steps:

- Understand your state's law
- Be transparent with anyone you give milk to about what it is and how it was handled
- Consider pasteurizing anything that leaves your property
- Talk to your insurer about whether farm liability coverage makes sense
- Keep records of testing and handling

None of this is legal advice. For anything involving money, a written agreement, or regular distribution to others, talk to an attorney who knows agricultural law in your state.

The practical summary

For yourself: do what you decide is right.

For others: learn the law first, and consider pasteurizing.

For sale: assume it is regulated, find out exactly how, and comply.

Chapter 22: Breeding, Drying Off, and Doing It Again

The cycle repeats. Here is how to run it.

Deciding when to breed

Work backward from when you want milk.

Goats: 5 months gestation. Breed in October, kid in March.

Cows: 9 months. Breed in June, calve in March.

Considerations:

- Avoid birthing in your worst weather. A newborn in an ice storm needs attention you may not want to give at 3 a.m.
- Match peak production to pasture if you want cheap milk - spring freshening
- Stagger multiple animals so one is always in milk
- Consider your own schedule. Birthing season demands attention

Cows are typically bred back 60 to 90 days after calving, which produces roughly a calf a year.

Goats are seasonal breeders in most breeds, cycling in fall for spring kids. Nigerian Dwarves and some others cycle year round.

How to breed

Artificial insemination (AI). A technician or you, with training, inseminates using frozen semen.

Advantages: no male to keep, access to superior genetics, no disease transmission, safer. Disadvantages:

requires accurate heat detection, timing is critical, and conception rates are lower than natural service for most beginners.

For cows, AI is usually the right answer for a homestead. Keeping a bull is genuinely dangerous - bulls injure and kill people every year - and expensive for one cow.

Natural service.

Own a male: convenient, and you are then feeding, housing, and containing a buck or bull year round for a few days of work. A buck in rut smells overwhelming and will taint your milk if he is anywhere near the does or the milking area.

Driveway breeding: take your female to someone else's male for a stud fee. Common and sensible for goats. Disease exposure is the trade-off - ask about the herd's testing status.

Lease a male for the season.

For goats, driveway breeding or a leased buck is usually better than owning one unless you have several does.

Detecting heat

You cannot breed her if you miss the cycle, and a missed heat costs 21 days.

Goat signs: tail flagging, vocalizing, restlessness, a swollen reddened vulva, discharge, mounting other does, decreased appetite, a drop in milk, and intense interest in a buck.

Cow signs: standing to be mounted is the definitive sign. Also restlessness, bellowing, clear stringy mucus, a swollen vulva, chin resting on other cows, and a drop in milk.

Write down every heat you observe. Two recorded cycles let you predict the next within a day.

A "buck rag" - a cloth rubbed on a buck in rut, kept in a sealed jar and offered to does - provokes an obvious reaction in a doe in heat. A crude and effective detection tool.

Confirming pregnancy

- No return to heat at the expected date. Suggestive, not conclusive
- Blood test, available through labs and quite accurate from about 30 days
- Milk test, similar
- Ultrasound, by a vet
- Palpation, for cattle, by a vet or an experienced person
- Visible changes later - udder development, belly shape, behavior

Confirm it. Assuming a pregnancy that does not exist means you dry her off for nothing and lose months of milk.

Drying off

Stopping lactation before the next birth. The udder needs to rest and repair, and the animal needs to put resources into the developing offspring.

When:

- Cow: 6 to 8 weeks before calving
- Goat: about 2 months before kidding

Do not skip it. An animal milked through to birth produces poorly in the next lactation and is at higher risk of metabolic problems.

Gradual method - most common:

1. Cut grain substantially. Less energy in, less milk out
2. Reduce milking frequency: twice a day -> once a day for several days -> every other day -> stop
3. Do not strip her out. Leave milk in the udder. Pressure is the signal that tells the udder to stop producing. Stripping her out tells it to keep going
4. Watch closely for mastitis during and after. The udder will be full and uncomfortable for several days, which is normal, and hot, hard, or painful is not
5. Once dry, check the udder daily for a week

Abrupt method: simply stop milking, all at once. Faster, more uncomfortable for the animal, and higher mastitis risk. Sometimes used with a veterinary dry-cow antibiotic treatment. Discuss with your vet, particularly for an animal with a mastitis history.

During the dry period:

- Good hay, minerals, water. Little or no grain
- Do not let her get fat - obesity causes hard births and metabolic disease
- Reintroduce a small amount of grain in the final 2 to 3 weeks - "steaming up" - to prepare the rumen for lactation feed
- Vaccinate on schedule so antibodies reach the colostrum
- Watch her condition

Freshening again

She gives birth. Colostrum for 2 to 3 days goes to the newborn.

Then you are milking again, and the cycle restarts.

Watch in the first days for:

- Milk fever - a calcium crash, seen as weakness, staggering, cold ears, collapse. An emergency, and dramatically responsive to prompt treatment
- Retained placenta beyond about 12 hours
- Udder edema - swelling, common and usually resolving
- Ketosis - an energy deficit in high producers, seen as reduced appetite and a sweet acetone smell on the breath
- Mastitis

Production climbs to peak over 4 to 8 weeks. Increase feed to match.

Culling

The part nobody enjoys.

Reasons to move an animal on:

- Chronic mastitis - she is a reservoir of infection for the rest
- Repeated failure to conceive
- Sharply declining production
- Dangerous or unmanageable temperament
- Chronic health problems
- Age

Options: sell her as a pet or a companion, sell her for meat, process her yourself (Book 11), or keep her as a retiree and accept the feed cost.

Decide in advance what you will do. People who have not thought about it end up with a field of unproductive animals eating hay, and that is how a small dairy quietly stops making sense.

Chapter 23: Your First Year

A month-by-month plan, and honest expectations.

Before you buy anything

[ ] Solve the vacation problem - identify and begin training a backup milker

[ ] Find a large animal vet who will take you as a client

[ ] Check local zoning and any HOA covenants

[ ] Build or buy shelter

[ ] Install fencing and test it

[ ] Build a milking stand or stanchion

[ ] Source hay and confirm storage

[ ] Buy minerals, buckets, and a water tank

[ ] Buy milking equipment (Ch. 8)

[ ] Find a mentor or a local dairy keeper who will answer questions

[ ] Watch someone milk. Better, milk under supervision

Do all of it before the animal arrives. Every one of these becomes an emergency if you skip it.

Months 1 and 2 - learning to milk

Goals: get competent at milking, get milk into the refrigerator clean and cold, establish a routine.

- Milk twice a day, same times
- Expect to be slow, sore, and clumsy for two weeks
- Focus on hygiene and chilling above everything. Those two habits set your quality permanently
- Strip check every teat every milking

- Do not attempt cheese yet. Drink it, and get comfortable
- Keep a daily record: yield, anything unusual

Expect: a production drop from the move, some frustration, and a genuine sense of accomplishment about three weeks in.

Months 3 and 4 - the kitchen

Goals: start converting surplus.

- Yogurt first. Easiest, most useful, hardest to ruin
- Then butter, if you have a cow. Get the cream temperature right
- Then a 30-minute cheese - ricotta or paneer
- Then chevre, which is where it gets interesting

Expect: more milk than you can drink. This is when you find out how much cheese you actually eat.

Months 5 through 7 - settling in

Goals: routine, confidence, and skill depth.

- Milking should now take half the time
- Try mozzarella
- Establish your surplus routine - what to do with milk on a given day, decided in advance
- Consider whether to move to once-a-day milking
- Watch body condition and adjust feed
- Get a fecal egg count done; learn FAMACHA if you have goats

Expect: this to feel normal. That is the point where it starts being enjoyable rather than effortful.

Months 8 through 10 - breeding and planning

Goals: set up next year.

- Breed, on the schedule you chose
- Confirm pregnancy
- Production is declining. Reduce grain accordingly
- Build a stock of frozen milk and aged cheese for the dry period
- Assess honestly: is this working? Is it one animal or three next year?

Months 11 and 12 - dry off

Goals: rest the animal, rest yourself.

- Dry her off on the schedule in Chapter 22
- Enjoy two months of not milking. It feels like a vacation
- Repair equipment, improve the milking area with everything you learned
- Prepare for freshening
- Read the cheesemaking book you have been meaning to

What to expect, honestly

You will:

- Be surprised at how much milk there is
- Get frustrated at the twice-a-day schedule around week three
- Spill a pail
- Make one genuinely bad cheese
- Have a health scare that turns out to be nothing

- Possibly have one that does not
- Get very good at milking without noticing it happened

You will not:

- Save money
- Get a vacation without planning for it
- Regret it, if you were suited to it in the first place

Scaling up, or not

After a year you will know.

Some people stop. They discover they resent the schedule, or the family does not drink that much milk, or the freedom mattered more than they thought. Selling the animal is a legitimate outcome, not a failure. Better to learn it in year one.

Some people add animals. Two does becomes four. Staggered freshening for year-round milk. Maybe a cow.

Some people go sideways - the same animals, deeper skills. Aged cheese, better genetics, showing, selling breeding stock.

The right size is the one you can sustain in February, in the dark, in the cold, when you have a head cold and it is your turn.

The reason to do it

There is a particular moment, maybe four months in, when the routine has stopped being a task.

It is early. Nobody else is up. The animal is on the stand eating grain and being entirely calm about your presence. Your hands know what to do and you are not thinking about it. Milk is hitting the pail in a steady rhythm, and it is the only sound.

Forty minutes later there is a cold jar of milk in the refrigerator that did not exist when you woke up, from an animal you know, on land you tend.

That is what people are actually after. It is not the milk, and the milk is very good.

QUICK REFERENCE

Production at a glance

Production at a glance

Animal	Daily milk	Butterfat
Holstein cow	6-10 gal	3.5%
Jersey cow	3-6 gal	4.5-5.5%
Dexter cow	1.5-2.5 gal	~4%
Saanen / Alpine goat	1-1.5 gal	3-3.5%
Nubian goat	0.75-1.25 gal	4.5-5%
Nigerian Dwarf goat	1-2 qt	6-10%

The cycle

The cycle

Method	Gestation	Lactation	Dry period
Cow	~283 days	~305 days	6-8 weeks
Goat	145-155 days	~305 days	~2 months

The cycle

Method	Heat cycle	Standing heat
Cow	~21 days	12-18 hr
Goat	18-21 days	12-36 hr

Daily feed, roughly

Daily feed, roughly

Method	Hay	Grain	Water
Goat	3-5 lbs	1-3 lbs	2-4 gal
Cow	25-40 lbs	4-12 lbs	20-30+ gal

Plus free-choice loose minerals. Goat minerals must contain copper. Never feed sheep minerals to goats.

The milking sequence

1. Wash hands

2. Animal on stand with grain
3. Brush off loose dirt
4. Wash udder and teats
5. Dry completely
6. Strip 2-3 squirts per teat into a strip cup - check for clots
7. Milk - trap, squeeze, release. Never pull
8. Milk out completely, strip the last
9. Dip every teat
10. Keep her standing 20-30 minutes - feed hay
11. Strain and chill immediately
12. Wash equipment

Equipment washing order

Cold rinse -> hot soapy wash -> periodic acid rinse -> clean rinse -> air dry inverted -> sanitize just before next use

Never hot water first - it cooks protein onto the surface.

Chilling

Target: below 40 degrees F within 30 minutes. Ice bath, then refrigerator. This is the single biggest factor in milk quality.

Chilling

Milk	Refrigerated life
Raw, chilled fast	7-10 days
Raw, chilled slowly	3-5 days
Home pasteurized	10-14 days

Pasteurizing

145 degrees F for 30 minutes (batch - preserves cheesemaking ability), then chill fast. 161 degrees F for 15 seconds (HTST). UHT milk cannot be used for cheese.

Cheese quick table

Cheese quick table

Cheese	Culture	Rennet	Time	Yield per gallon
Ricotta	No	No	30 min	1 1/2-2 cups
Paneer	No	No	45 min	~1 lb
Mascarpone	No	No	12-24 hr	-
Chevre	Yes	1 drop	12-18 hr + drain	~1 1/2 lb
Cream cheese	Yes	1 drop	12-18 hr + drain	-
Mozzarella	No (citric acid)	Yes	30 min	~1 lb
Farmhouse hard	Yes	Yes	Hours + 2-12 months	~1 lb per gal

Butter

Cream at 55-60 degrees F. Too cold will not break; too warm goes greasy. Wash until the water runs clear, or it goes rancid. ~2 cups cream -> ~8 oz butter + ~1 cup buttermilk.

Yogurt

180 degrees F -> cool to 110-115 degrees F -> culture -> hold 105-110 degrees F for 4-12 hours -> chill 4 hours.

Mastitis - check every milking

Signs: clots or flakes in the milk - hot, hard, swollen quarter - watery or bloody milk - sudden production drop - kicking

Do: call the vet - milk that quarter out frequently and completely - discard the milk - observe withdrawal exactly

Prevent: dry bedding - clean dry teats - complete milk-out - post-dip every teat every milking - keep her standing 20-30 min after

Key contacts

Before the animal arrives, keep these contacts somewhere easy to reach: your veterinarian, after-hours vet, backup milker, experienced local mentor, county extension office, and state department of agriculture.

GLOSSARY

Acid whey - Whey from acid-set or cultured products. Around pH 4.5. Limited uses.

Butter muslin - Fine-weave cheesecloth. Grocery-store cheesecloth is too open for cheesemaking.

CAE - Caprine arthritis encephalitis. A chronic incurable goat virus. Test before buying.

Chevre - Fresh soft goat cheese, culture-set.

Clean break - The test for a properly set curd: a knife lifted through it splits it cleanly with clear whey filling the cut.

CL - Caseous lymphadenitis. A contagious abscess disease of goats and sheep. Never leaves a property.

CMT - California Mastitis Test. A paddle test for subclinical mastitis.

Colostrum - The first milk after birth. Antibody-rich, and it belongs to the newborn.

Culling - Removing an animal from the herd.

Drying off - Ending lactation before the next birth.

FAMACHA - Eyelid color scoring for anemia from parasites in goats and sheep.

Freshening - Giving birth and beginning lactation.

Ghee - Clarified butter with the milk solids browned and removed. Shelf stable.

Let-down - The oxytocin-driven release of milk from the udder into the teats.

Mastitis - Inflammation of the udder, usually bacterial. The main dairy disease.

Mesophilic culture - Bacteria working at room temperature. Used in chevre, cheddar, most cheeses.

Milk fever - A calcium crash around freshening. An emergency.

Pasteurization - Heating milk to destroy pathogens. 145 degrees F/30 min or 161 degrees F/15 sec.

Rennet - An enzyme that coagulates milk protein. Animal, microbial, or vegetable.

Somatic cell count - A measure of immune cells in milk. High counts indicate mastitis.

Steaming up - Feeding grain in the last weeks before freshening to prepare the rumen.

Sweet whey - Whey from rennet-set cheese. Around pH 6. The useful kind.

Thermophilic culture - Bacteria working at higher temperatures. Mozzarella, Parmesan, yogurt.

UHT - Ultra-pasteurized. Cannot be made into cheese.

Wether - A castrated male goat.

Withdrawal period - The time after treatment during which milk must be discarded.

WHERE TO GET HELP

A large animal veterinarian. Establish the relationship before you need it. Many practices will not take emergency calls from non-clients.

A local mentor. Someone keeping dairy animals in your climate. Worth more than any book, because feed, parasites, and seasons are all local.

Cooperative Extension. Every state has one. Many publish dairy goat and family cow guides, run FAMACHA certification, and provide fecal testing. Search your state plus "extension dairy goats" or "extension family cow."

Breed associations. ADGA (American Dairy Goat Association), AGS (American Goat Society), and the various cattle breed registries. Breeder directories, standards, and production records.

Your state department of agriculture. For raw milk law, licensing, animal identification, and testing requirements. The authority on anything legal.

Books worth owning:

- Storey's Guide to Raising Dairy Goats, Jerry Belanger and Sara Thomson Bredesen - the standard goat reference
- Keeping a Family Cow, Joann Grohman - the classic on the family cow, opinionated and useful
- Home Cheese Making, Ricki Carroll - the beginner cheesemaking standard
- Mastering Artisan Cheesemaking, Gianaclis Caldwell - the serious next step
- The Art of Natural Cheesemaking, David Asher - a traditional, low-input approach

Cheesemaking suppliers. New England Cheesemaking Supply and similar carry cultures, rennet, butter muslin, and molds, and most publish free recipes that are more reliable than what you will find on a blog.

Online communities. Dairy goat and family cow forums and groups are active and generous. Useful for troubleshooting, and cross-check medical advice with your vet.

Be careful with: anyone claiming raw milk carries no risk, anyone claiming it is uniquely dangerous, sources recommending blanket scheduled deworming, and advice about seasons and feed from a climate unlike yours.

ABOUT THE AUTHOR

Hudson Briggs writes practical, beginner-friendly guides for people who want to build useful skills at home. His books focus on simple steps, realistic numbers, and the confidence that comes from making a more capable household one project at a time.